SCIENCE EXPERIMENTS for CHILDREN

in 5 Volumes

Vol.- 1

249 Experiments on PHYSICS

V. Khatri

PUSTAK MAHAL®

Publishers
Pustak Mahal®

Administrative office and sale centre

J-3/16 , Daryaganj, New Delhi-110002
☎ 23276539, 23272783, 23272784 • ***Fax:*** 011-23260518
E-mail: info@pustakmahal.com • ***Website:*** www.pustakmahal.com

Branches
Bengaluru: ☎ 080-22234025 • ***Telefax:*** 080-22240209
E-mail: pustakmahalblr@gmail.com
Mumbai: ☎ 022-22010941, 022-22053387
E-mail: unicornbooksmumbai@gmail.com

ISBN 978-81-223-1620-9

Edition: 2022

Printed at: Radha Offset, Shahdara, Delhi - 93

Contents

1. Ray Marker

In this experiment, we will make a simple device that produces side-by-side light beams. By observing the behaviour of these beams, you'll be able to learn more about the properties of light.

Material Required

- Index card;
- Comb;
- Pair of scissors;
- Tape;
- Flashlight;
- Mirror;
- Clay.

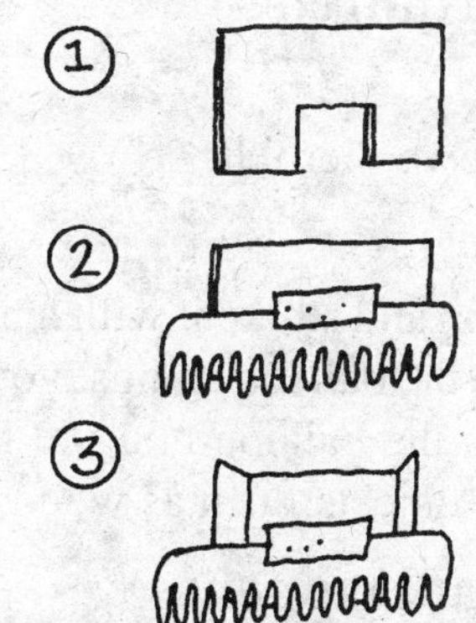

Procedure

To construct a "ray maker," cut a rectangular opening in the centre of an index card's long edge. Use tape to secure a comb over this opening. Try to position the thicker comb's teeth in the opening. Bend back the side edges of the card so that it stands freely.

Secure a mirror with a small lump of clay. Set the upright mirror in the centre of a table. Place the ray maker several inches from the mirror.

Flip on your flashlight. Aim the light beam through the teeth of the comb so that the light strikes the mirror. Observe what happens to these beams as they reflect off the mirror's surface. Do the beams change direction? How does the angle at which they strike the mirror compare to the angle at which they reflect off the mirror's surface?

This is What Happens

Light that travelled through the openings between the comb teeth formed visible beams. These beams struck the surface of the mirror and were reflected. By examining the beams and their reflections, you can uncover how light behaves when it strikes a mirror.

Magical Word: Save this ray-making tool! You'll need it for several other experiments.

2. Glued Balloons

Birthday parties occur throughout the year. Most likely, some of the parties you've attended had balloon decorations. By rubbing the balloons and placing them against the wall, they would magically cling to the dry flat surface. Why?

Material Required

- Balloon;
- A piece of wool or felt.

Procedure

Inflate a balloon and stroke it with a piece of wool or felt. If you don't have any fabric material, you can stroke the balloon against your hair. Place the balloon against the wall. What happens? How long will the balloon cling to the wall? Recharge the balloon and observe how well it clings to other objects, such as wood, a metal cabinet and glass.

This is What Happens

As the balloon is rubbed with wool, it becomes negatively charged. This charge produces an invisible electric field.

When the balloon is brought close to the wall, the negative charges in the wall are repelled by the balloon's negative field. These charges move deeper (migrate) into the wall. As they leave the wall region nearest the balloon, they create a positive wall surface. As a result, the positive wall and the negative balloon attract.

The wall became charged by induction. In this type of charging, objects don't touch. Instead, an electric field causes charges to migrate in surrounding materials. Although the number of charges remain the same, they are unequally distributed. Regions with more positive charges take on a net positive charge. Regions with more negative charges take on a net negative charge.

3. Light with Electricity

We all have tube lights everywhere around us, in our home, schools, etc. But have you ever wondered how a fluorescent tube (light bulb) works.

Material Required

- Balloon;
- Fluorescent tube.

Procedure

Inflate and tie the balloon. Wash the outside of the fluorescent tube and thoroughly dry. in a dark room, place one end of the tube against the floor. Hold the tube upright and quickly rub the balloon up and down the outside of it. Hold the balloon near the tube.

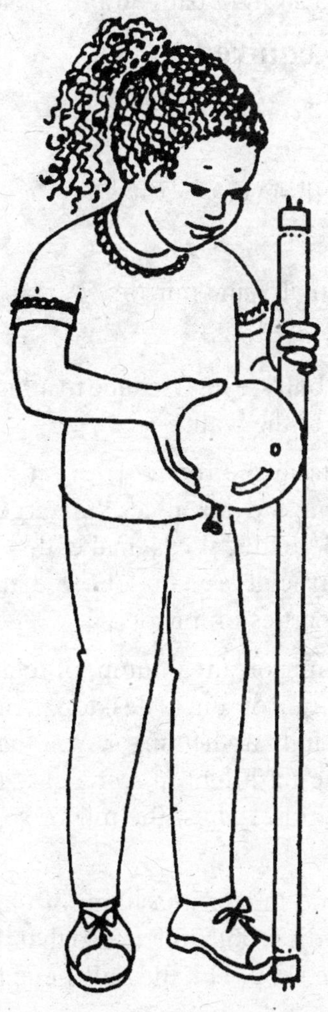

This is What Happens

The fluorescent tube starts to glow and the light moves with the movement of the balloon. Once the tube starts glowing, even the nearness of the balloon causes light to be produced.

Science Behind It

When a fluorescent tube is connected to an electrical current, the chemicals on the tiny filaments at each end of the tube produce electricity that can jump from one end of the tube to the other, producing 120 flashes of light every second. This light produces ultraviolet light—rays invisible to the human eye. A drop of mercury inside the tube is vaporized by the electrical flashes and the vapour carries electrons to the phosphor powder coating the inside of the tube. This coating changes the ultraviolet energy into light energy that can be seen. Rubbing the balloon on the tube causes the same changes to occur, but on a smaller scale. Rubbing the balloon causes electrons to build up on the surface of the balloon. This buildup of electrons causes the mercury vapour inside the tube to become charged, and just as when the tube is connected to an electric current, the charged mercury vapour bombards the fluorescent chemicals, resulting in visible light.

4. Bouncing Beams

Most of the light that strikes a mirrored surface is not absorbed by the material. Instead, this light bounces off the reflective surface. Since the mirror is flat, the light rays are not randomly scattered. Instead, they travel together side-by-side as a reflected beam. The beam keeps travelling on a straight path until it strikes another surface.

Material Required

- Flashlight;
- Clay;
- Target bull's-eye;
- Tape;
- Several small hand mirrors.

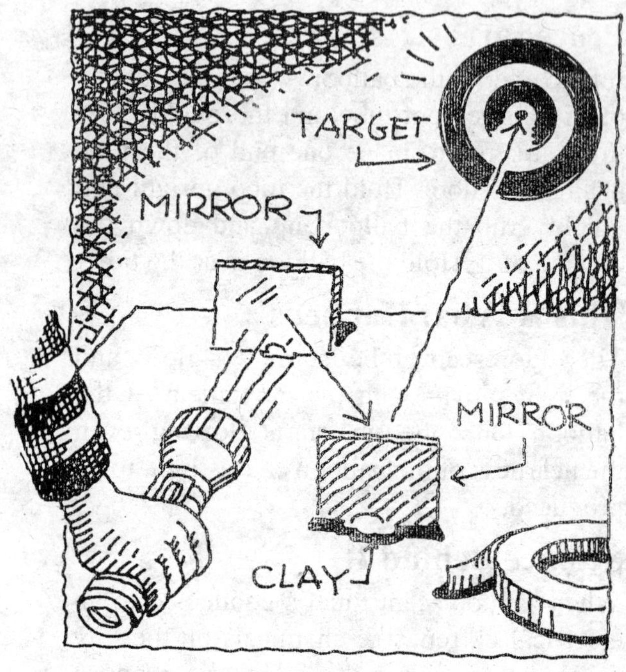

Procedure

Draw a target bull's-eye. Use tape to attach the bull's-eye to the wall.

Dim the lights in the room. Turn on the flashlight. Aim the beam at the target. Not too hard to do? Let's make this a little more difficult as you observe the reflective properties of mirrors.

Set a small mirror in a lump of clay. Position this mirror on a desktop. Aim your beam of light at the mirror. You may have to set the flashlight in some clay to secure its position. Adjust the mirror so that the flashlight's beam is reflected upon the target. Good.

Set up a second mirror in a lump of clay. Position this mirror a few feet away from the first mirror. Now adjust both mirrors so that the beam bounces from the first to the second. The final reflected beam should hit the bull's-eye target.

How many more mirrors can you add to this setup?

This is What Happens

Light that struck the mirror's surface was reflected in another direction. This reflected beam struck another mirror and was reflected again. Eventually, the beam was projected onto the target. The target surface was not a mirror. It did, however, reflect light. This reflected "spot" was detected by your eyes.

5. Contact Changing

From the previous experiment, we made negative charges transfer onto a balloon when it was rubbed with wool. Because the balloon now has more negative charges than positive charges, it has a net negative charge. This type of "charging" is called contact charging.

Contact charging can also produce objects with a positive charge. For an object to become positive, it must lose some of its negative charges. This loss offsets the neutral balance to produce a net positive charge.

Material Required

- 10 inch strip of nylon fabric;
- A pair of scissors;
- Plastic grocery bag.

Procedure

Use your scissors to cut a 10-inch strip of nylon. Firmly hold the middle of the strip, allowing the halves to hang downward. Grasp the nylon with the plastic bag. Stroke both halves of the fabric several times. What happens when you stop stroking the nylon? What causes the nylon to behave this way?

This is What Happens

Unlike wool, your plastic does not easily give up its negative charges. On the contrary, it has a tendency to take in negative charges. When the plastic rubbed along the nylon, negative charges are transferred to the plastic material. This left the nylon strips with a positive charge. Since both hanging halves had the same charge, they repelled each other, causing the free ends to separate.

6. Good and Bad Conductor

Do all materials conduct electricity? Let's find out in this simple experiment.

Material Required

- Spring-type clothespin;
- 1 D-cell battery;
- Aluminum foil;
- Flashlight bulb;
- Masking tape;
- Scissors;
- Testing materials: rubber band, paper, coins ruler.

Procedure

Cut a rectangle of aluminum foil, 24 in. x 12 in. (60 cm x 30 cm). Fold the aluminum piece in half lengthwise five times to form a thin strip 24 in. (60 cm) long. Cut the aluminum strip in half to form two 24 in. (30-cm) strips. Tape one end of each of the metal strips to the ends of the battery.

Wrap the free end of one of the metal strips around the base of the flashlight bulb. Hold the tape in place with the clothespin.

Test the electrical conductivity of the materials collected by touching the metal tip on the bottom of the flashlight bulb to one side of the material while touching the free end of the metal strip to the opposite side of the same material.

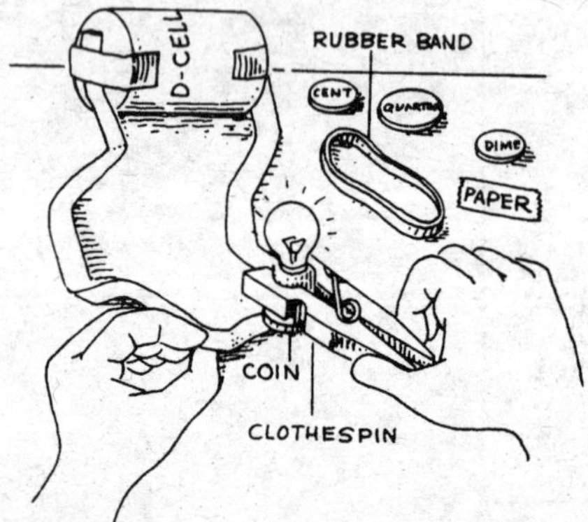

This is What Happens

The coins were the only materials that caused the bulb to glow.

Science Behind It

An electric circuit is the path through which electrons move. A switch is a material that acts as a bridge or pathway for the electrons. When the, switch is closed, the electrons move freely, but when it is open, the electrons stop. The only materials tested that allowed electrons to flow through them were those made of metal. Touching the paper clip to one side of a piece of metal and the tip of the bulb to the other side allowed the electrons to flow out of the negative part of the battery through the aluminum tape (conductor) into the bulb. The electrons continue their path from the bulb through the aluminum strip and back into the positive end of the battery. As long as there is no break in the system, the electrons continue to flow and the bulb will continue to glow.

7. Ways of Ray

Does shooting a billiard ball against the side bumper of a pool table has same effect as a light ray? Find out in this experiment.

Material Required

- Flashlight;
- Small mirror;
- A ruler;
- Tape.

Procedure

Use tape to place a set of marks along one wall of a hallway or narrow room. Place the first mark 1 foot above the floor. Place the second mark 1 foot above the first (2 feet high). Add two more marks, one at 3 feet high and the final one at 4 feet high. Make sure that when you position the marks, they are placed directly atop each other.

Place a mirror in the centre of the hallway floor. The mirror should be placed in line with the set of wall marks.

Dim the lights in the hallway. Position your flashlight at the 1-foot mark. The back end of the flashlight should remain in contact with the wall. Aim the beam at the mirror. Observe where the reflected spot appears. Mark this spot with a piece of tape.

Position the flashlight at the other marks along this wall. Observe and mark where the reflected spot appears. How does the angle at which the flashlight beam strikes the mirror affect its reflection?

Suppose the beam came from directly above the mirror. Where would the reflected spot appear?

This is What Happens

The angle at which the beam strikes the mirror determines the angle of reflection. If the striking beam (also called the incident beam) hits this surface at a shallow angle, the reflected beam will travel outward along a shallow angle. Likewise, if the incident beam strikes the mirror at a steep angle, the reflected angle will also be steep.

In your hallway experiment, the spot always appeared on the opposite wall at the same height from which the beam began. Since the angle of the incident ray equalled the angle of the reflected ray, this spot had to appear at the same height.

8. Changing Charges

So far you've observed what happens when objects of the same charge are brought together. The negative balloons repel each other. The positive nylon strips repel each other. But what happens when a negative balloon and a positive strip are brought near each other? Find out in this easy experiment.

Material Required

- Nylon strip;
- 1-foot thread;
- Tape;
- Balloon;
- Wool;
- Plastic grocery bag.

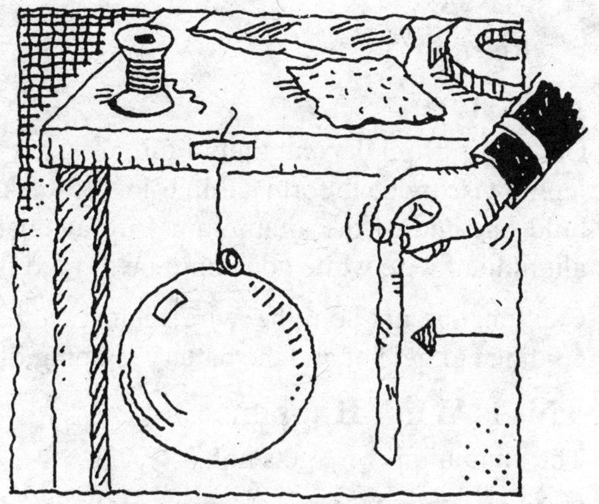

Procedure

Attach a 1-foot length of thread to an inflated balloon. Charge the balloon by rubbing it with wool. Use tape to attach the thread to the edge of a table or desk.

Charge the nylon strip by rubbing it with a plastic bag. Stroke the strip several times to ensure that it becomes sufficiently charged. Bring the strip near the hanging balloon. What happens?

Release the strip. Does it stick to the balloon or is it repelled by this oppositely-charged object?

This is What Happens

Like charges repel. Unlike charges attract.

Both the balloon and strip were charged by contact charging. The balloon took on a net negative charge. The nylon strip took on a net positive charge.

When the negative and positive charges were close enough, the objects moved together. At close range, the attraction was strong enough to stick the nylon to the balloon's surface.

9. Hot Electrons

The purpose of this exciting experiment is to discover that the flow of electrons generates heat.

Material Required

- 1 AA battery;
- Aluminium foil;
- Scissors;
- Ruler.

Procedure

Cut a strip from the aluminium foil, 6 in. x 1 in. (15 cm x 2.5 cm). Fold the strip of foil in half lengthwise twice to form a thin 6-in. (15-cm) strip that will be used as a wire. Using one hand, hold one end of the aluminium wire against each battery pole. After 10 seconds, touch the aluminium wire while you continue to hold the wire against the battery ends.

Caution: Do not hold the wire against the battery ends longer than 20 seconds. The wire will continue to get hot and the battery is being discharged (losing its power).

This is What Happens

The aluminium wire gets hot.

Science Behind It

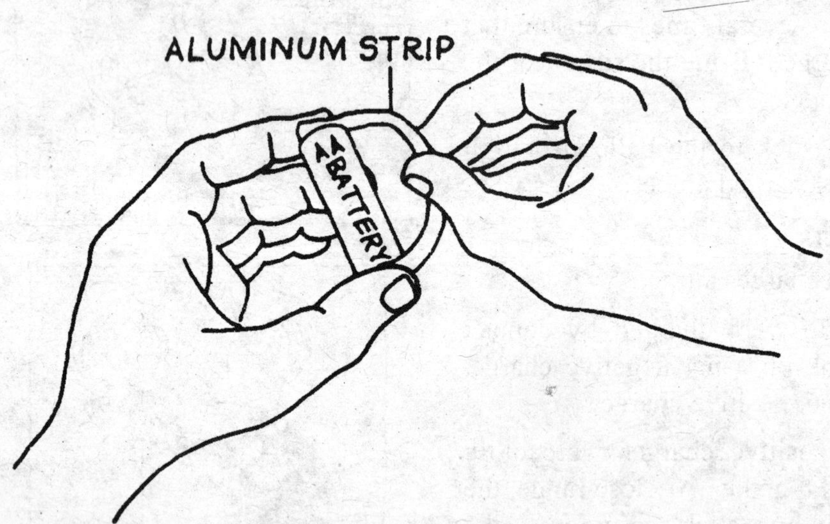

Touching the wire to the ends of the battery produces a path through which electrons travel (an electric circuit). The electrons move out of the negative end of the battery through the wire and back into the positive end of the battery. The movement of the electrons causes the wire to get hot.

When a light bulb is placed in the electrical circuit, the electrons move through the bulb. The movement of the electrons heats up the wire filament inside the bulb. The hot wire filament becomes incandescent, that is, it gets so hot that it gives off light.

10. Reflected Beams

Can you imagine life without a TV remote control? Terrible! But believe it or not, prior to the 1970s most television sets did not have remote control. In order to change the channel, a person had to stand up, walk to the TV, and manually turn the knob.

Material Required

- TV with remote control;
- Mirror.

Procedure

Switch on the television. Now use your remote control device to change channels. Determine how far away from the TV the remote can control it. Place a sheet of paper between the remote and the television. Does the remote still work or does the paper block the signal?

Stand about 10 feet from the television. Aim the remote at the side of the TV. How far away from the TV can the remote be aimed before it no longer switches channels?

Can the beam bounce off the ceiling and still reach the TV? Try reflecting the beam off the ceiling and wall. Will the beam be reflected off or absorbed by these surfaces?

Place a mirror on the floor about midway between you and the television. Hold the remote at the same height as the remote receiver on the television set. Aim the beam at the mirror. What happens? Can you explain your observation?

This is What Happens

Although the remote control doesn't use visible light, it uses a similar form of energy called infrared light. Infrared, also called IR, has many of the properties of visible light. Like the colours we see, IR can bounce off mirrored surfaces. That's why the beam that struck the mirror changed the channel. Like visible light, it was reflected from the surface and travelled along a path that ended at the TV remote receiver.

11. Invisible Legs

The magician covers a box with a silk scarf. Slowly, she raises the box into the air. While holding the covered box, she suddenly flicks her wrist. The scarf goes limp and the box disappears. How did she do it? We're not telling, but we do have a magical activity that also produces a phantom shape in thin air.

Material Required

- Sheer nylon stocking;
- Plastic grocery bag;
- Smooth wall.

Procedure

With one hand, hold the top of the stocking against a flat wall. Use the other hand to stroke the stocking in one direction with the plastic grocery bag. As you stroke the nylon material, smooth it against the wall's surface.

After several strokes, release the stocking. What happens to the material? How is this similar to sticking balloons to a wall? How is it different?

Now gently grasp the top of the stocking and move it away from the wall. Make sure that it does not touch anything (including yourself). Hold out the stocking. What happens to its shape? Can you explain your observation?

This is What Happens

As the plastic bag moves over the nylon, it picks up negative charges. This produces a stocking with a net positive charge. This charged stocking acts like the negatively-charged balloon and induces an opposite charge in the nearby wall. The negative and positive charges attract and the stocking clings to the wall. When the stocking is pulled from the wall, it still retains its net positive charge. These charges, which are distributed throughout the stocking material, push away from each other. This causes the stocking to "expand" and take on the leg shape it was manufactured in.

12. Making a Galvanometer

Galvanometer is an instrument for measuring small electric currents by deflection of a moving coil. In this experiment, we will make one and determine if an electric current affects a magnet.

Material Required

- Aluminum foil, 1 yd. (1 m);
- Compass;
- Cardboard box to fit the compass;
- Scissors;
- D-cell battery.

Procedure

Cut a piece of aluminum about 36 in. x 24 in. (100 cm x 60 cm). Fold the aluminum piece in half lengthwise five times to form a thin strip 36 in. (100 cm) long.

Place the compass in the box. Wrap the aluminum strip around the box as many times as possible, leaving about 6 in. (15 cm) free on both ends of the strip. Turn the box with the compass so that the aluminum winding is pointing in a north-to-south direction. Tape one end of the aluminum strip to the positive end of the battery.

Watch the compass needle as you touch the free end of the metal strip to the negative end of the battery. Alternately touch the metal strip to the battery and remove it several times.

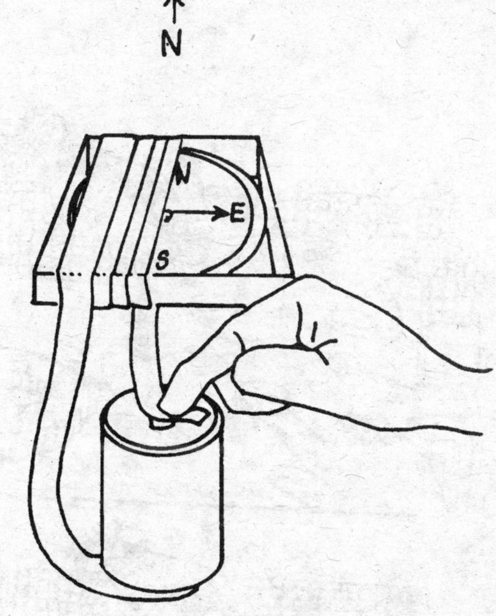

This is What Happens

The needle on the compass moves away from, and then returns to, its north-to-south direction when the metal strip is touched to and then removed from the battery.

Science Behind It

Electrons flow out of the battery, through the aluminum strip, and back to the battery. Moving electrons produce a magnetic field. Since the aluminum strip is turned in a north-to-south direction, the movement of electrons through the strip produces a magnetic field pointing east and west. The needle of the compass will be pulled towards this magnetic field, thus indicating that an electric current is flowing through the strip. The larger the current through the strip, the stronger the magnetic field that is produced.

13. Hall of Mirrors

Reflections can be confusing, especially when you are trying to find your way out of a funhouse in Melas or Circus. No matter where you look, a never-ending series of reflections look back. Only by feeling for an opening can you navigate through this maze of mirrors. In this next experi-ment, you'll build a much smaller, but equally confusing mirror hall.

Material Required

- Three small plastic mirrors (with frames removed);
- Coin;
- Clay;
- Pencil;
- Tape.

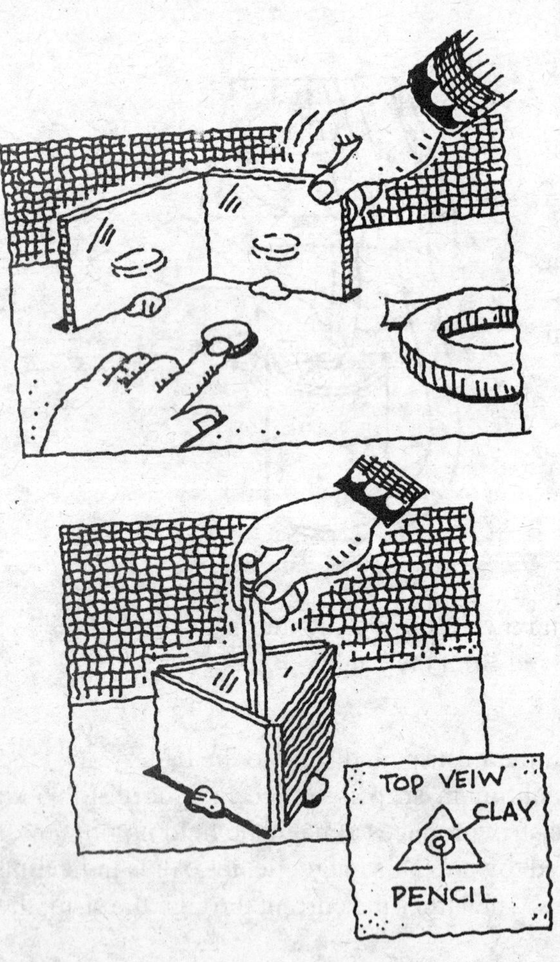

Procedure

Set the base of each mirror in a small lump of clay. Arrange two mirrors on a flat surface so that their sides form a right angle. Place a coin in the centre of these two mirrors. Examine the coin's reflection in the mirrors. Change the angle at which the mirrors connect. How does this affect the reflected image?

Remove the coin. Add a third mirror so that a triangle is formed. Stand a pencil in a clay base. Position the pencil in the centre of this tiny hall of mirrors. Peer into the middle and what do you see?

This is What Happens

It's that mirror thing again. When you look into one of the mirrors, you see more than the object. You observe a reflection that is reflected and reflected and reflected. This multiple reflection produces a series of images that can appear endless.

14. Rolling Barrel

So far, you've learned that paper sticks well to both combs and walls. But did you know you can move a rolled-up piece of paper by static attraction? Here's how.

Material Required

- Paper;
- Pair of scissors;
- Tape;
- Comb;
- Wool.

Procedure

Cut out a strip of paper about 1 inch wide and 4 inches long. Roll the paper into a cylinder and secure its shape with a small section of tape. Charge up the comb by running it through your hair or stroking it with a piece of wool or flannel. Place the comb near the paper barrel. What happens?

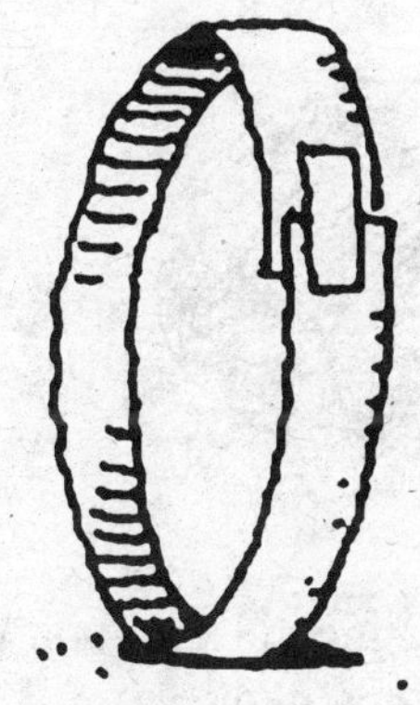

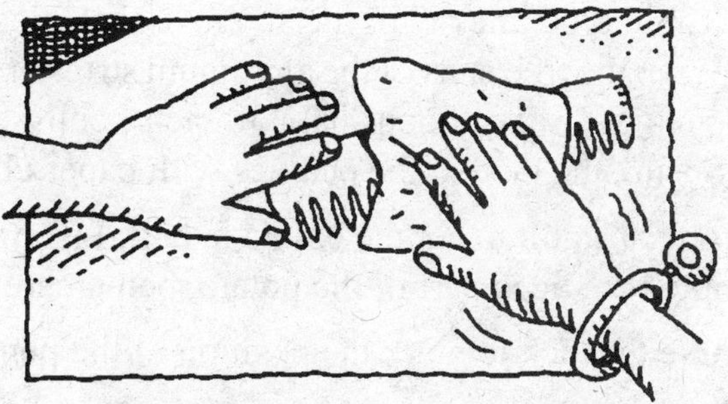

This is What Happens

Stroking the comb gives it a net negative charge. As it was brought near the neutral-charged paper barrel, it induced a charge in the paper. Negative charges on the closer side of the paper were repelled away, leaving a net positive region. This region was attracted to the comb. When inertia and friction were overcome, the paper rolled towards the comb.

15. Potato Positive Terminal Taster

The aim of this easy experiment is to determine which battery terminal is positive.

Material Required

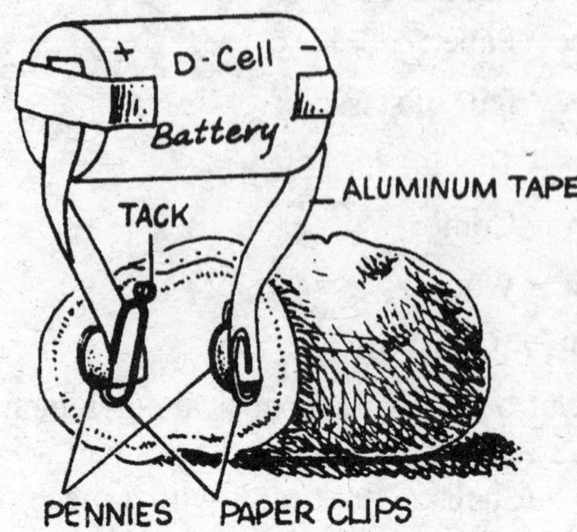

- Aluminum foil;
- D-cell battery;
- Potato;
- 2 paper clips;
- 2 pennies;
- Steel wool;
- Masking tape;
- Scissors;
- 1 push tack.

Procedure

Cut a piece of aluminum foil 24 in. x 12 in. (60 cm x 30 cm). Fold the aluminum piece in half lengthwise five times to form a thin strip 24 in. (60 cm) long. Cut the aluminum strip in half to form two 24 in. (30 cm) strips.

Rub the pennies with the steel wool to clean them. Wrap the free ends of the aluminum strips around the pennies. Leave about half of each penny exposed. Secure the tape to the pennies with paper clips.

Ask an adult to cut the potato in half. Insert the edge of the pennies about 1/2 inch (1 cm) apart into the sliced part of the potato, being careful to keep the aluminum strips attached.

Use the tack to mark the position of the penny connected to the positive end of the battery.

Remove the pennies after one hour. Examine the holes made by the pennies.

This is What Happens

The hole around the penny connected to the metal tape leading to the positive terminal is green.

Science Behind It

Connecting the penny to the positive pole of the battery gives the copper in the penny a positive charge. When the positively charged copper particles combine with negative particles in the potato, a green copper compound is formed. You can use this experiment to identify the positive terminal of any battery.

16. Making Kaleidoscope

A kaleidoscope is a toy that uses reflected images to produce great patterns. Would you like to see how one works? Just follow these instructions.

Material Required

- Three small plastic mirrors (with frames removed);
- Tape;
- Wax paper;
- Pair of scissors;
- Rubber band;
- Several small and transparent coloured beads.

Procedure

Arrange the mirrors to form a triangular viewing tube. The reflecting sides of the mirrors should face inward. Use tape to secure this shape. Cut out a piece of wax paper that is slightly larger than the triangular open end. Cover one end with the wax paper. Pull the paper tight. Use a rubber band to secure the paper to the tube end. Drop several small coloured beads into the tube. If you don't have beads, small pieces of coloured plastic or tissue paper will also work. Look into the open end of the kaleidoscope. What do you see? How many reflections are there? Shake the tube so that the beads move to new positions. What do you see now?

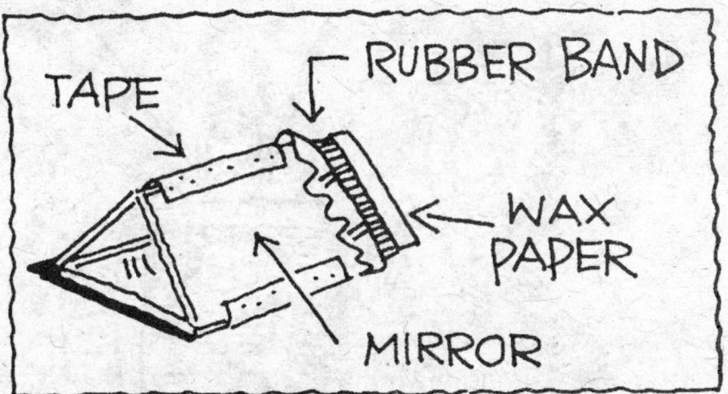

This is What Happens

The magic of the kaleidoscope arises from its creation of multiple reflections. The angle of the three mirrors, however, limits the number of reflections. Each of the reflections (and the real screen) appears attached to its neighbouring reflections.

17. Crackles and Pops

Here is a net experiment that you can perform using your television set.

Material Required

- Television set or computer monitor;
- A sheet of paper.

Procedure

Switch the television set on. Roll up your shirt sleeve. Place your forearm against the television screen. Hear anything? Feel anything? Slowly move your arm across the screen. What happens now?

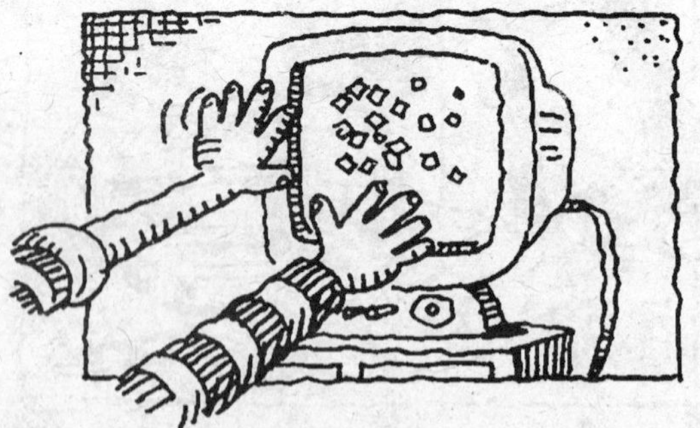

Tear up a sheet of paper into small fingernail-sized pieces. Put these pieces on the television screen and take your hand away. Do the pieces fall to the floor? Slowly move your hand just above these pieces of paper. Do any of the pieces react? If so, how? Touch several of them. Does this change their behaviour?

This is What Happens

The television screen is a charged surface. As your arm moved along the screen, it entered the screen's electric field. This field induced a charge on your arm hair so that they became attracted to the screen. As a result, you felt them standing on end! As the hair approached the screen, the static charge produced a small jump of electricity that made a crackling sound.

Like your arm hair, the pieces of paper became charged by the screen's electric field. The paper's charge caused it to stick to the screen. When your hand moved above the paper, the charges shifted again to cause the paper to fly off the screen.

18. Moving Paper

In this experiment, we will use simple material with is easily available in every home to charge an object with static electricity.

Material Required

- Comb;
- Tissue paper;
- Scissors;
- Ruler.

Procedure

Cut a strip of tissue paper about 3 in. x 10 in. (7.5 cm x 25 cm). Cut long, thin strips in the paper, leaving one end uncut (see diagram). Quickly move the comb through your hair several times. Your hair must be clean, dry, and oil-free. Hold the teeth of the comb near, but not touching, the cut end of the paper strips.

This is What Happens

The thin paper strips move towards the comb.

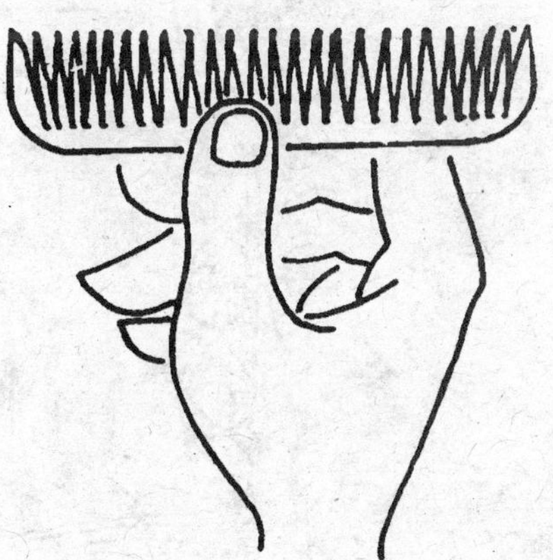

Science Behind It

Static means stationary. Static electricity is the buildup of negative or positive stationary charges. Matter is made up of atoms, which have electrons spinning around a positive centre called the nucleus. Moving the comb through your hair actually rubs electrons off the hair and onto the comb. The side of the comb that touched your hair has a build-up of electrons, making that side negatively charged.

The paper strip is made of atoms. Holding the negatively charged comb close to the paper causes the positive part of the atoms in the paper to be attracted to the comb. This attraction between negative and positive charges is strong enough to lift individual strands of paper.

19. Magical Spoon

Fun house mirrors can reflect some pretty strange images. Look into one mirror and your body might appear 10 feet tall. Walk in front of another mirror and you might look as if you didn't have a midsection. These kinds of distortion are produced by curves.

Material Required

- Shiny metal tablespoon.

Procedure

Hold up a shiny metal tablespoon. Note that the bowl-like part of the spoon has two sides. The side of the bowl that bulges outward is called the convex side. When we use the spoon to eat, the convex side faces downward. The side of the bowl that goes inward (forming a depression) is called the concave side.

Hold the spoon vertically. Look into the convex side of the spoon. How does your image appear? Is your image right-side up or is it flipped upside down? Is your image stretched? If so, do you look taller or wider?

Now hold the spoon horizontally. Look into the convex side again. How does the distortion change as you turn the spoon?

Hold the spoon vertically. Flip it around so that you are looking at the concave side. How does your image look now? Is it right-side up? Are your features distorted?

Hold the spoon horizontally. How does it affect the image now? Slowly bring the spoon closer to your face. Does the image flip or does it remain the same?

This is What Happens

The distortion is caused by the curved surface of the mirror. As side-by-side light rays strike the curves, they are sent back at slightly different angles. It's this spreading or concentrating of rays that causes the reflections to take on odd appearances.

20. Bending Water

"Never use an electric appliance while you are in the bathtub!"

Although it is not a great conductor of electricity, water can easily conduct the current that flows from your home's power outlet. The results are often deadly.

What about static electricity? Can water interact with non-moving charges?

Material Required

- Plastic comb;
- Pencil;
- Plastic pen;
- A piece of wool;
- Sink.

Procedure

Turn on a water faucet. Adjust the flow to a slow but steady stream. Pass the plastic comb through your hair several times. Slowly bring the comb close to the water. What happens?

Stroke a plastic pen with a piece of wool material. Now bring the pen towards the water. What happens?

Repeat this activity using a pencil. Does the pencil produce the same effect as the pen? Why?

This is What Happens

Plastic is a good material for storing electric charges. As the comb travelled through your hair, it picked up a negative charge. When it was brought close to the running water, it induced a positive charge in the closest part of the flow. The positive water and the negative comb attracted and produced a bend in the flow.

As you might have guessed, wool is a poor storage material for an electric charge. It did not keep enough charge to affect the water stream.

21. Making Simple Electroscope

An electroscope is used to detect the presence of electric charges, to determine whether these charges are positive or negative, and to measure and indicate their intensity. By the help of this electroscope, we will see the movement of electrons.

Material Required

- Aluminum foil, as thin as possible;
- Glass jar, 1 qt. (1 litre);
- Soft plastic lid to cover top of jar;
- Large paper clip;
- Needle-nosed pliers;
- Modeling clay;
- scissors Balloon, 9 in. (23 cm);
- Pencil;
- Scissors;
- Adult helper.

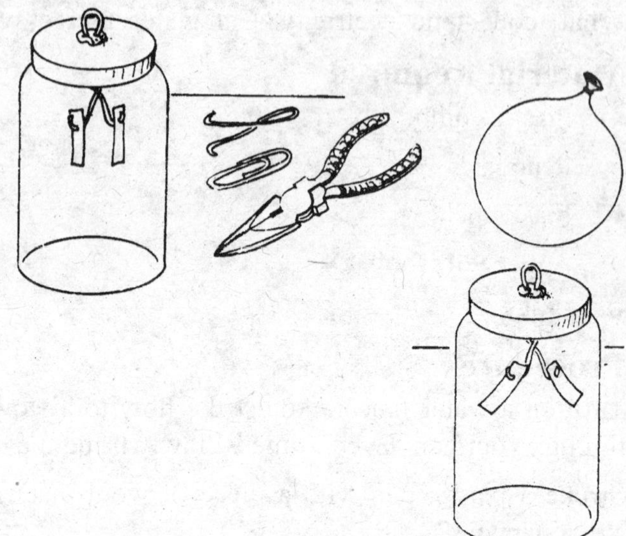

Procedure

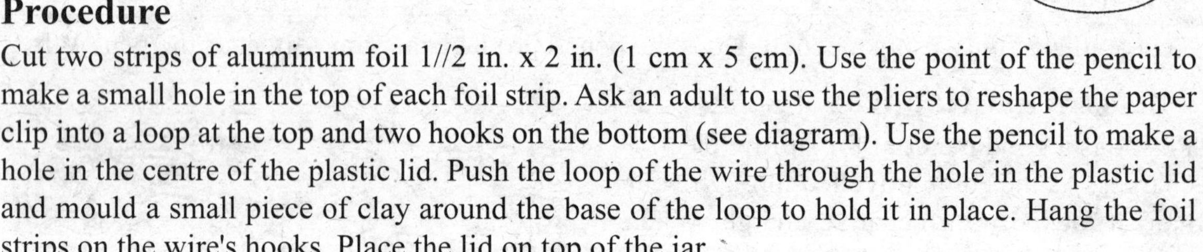

Cut two strips of aluminum foil 1//2 in. x 2 in. (1 cm x 5 cm). Use the point of the pencil to make a small hole in the top of each foil strip. Ask an adult to use the pliers to reshape the paper clip into a loop at the top and two hooks on the bottom (see diagram). Use the pencil to make a hole in the centre of the plastic lid. Push the loop of the wire through the hole in the plastic lid and mould a small piece of clay around the base of the loop to hold it in place. Hang the foil strips on the wire's hooks. Place the lid on top of the jar.

Inflate the balloon and rub it on your hair. Your hair must be clean, dry, and oil-free. Hold the balloon near the loop, on top of the jar.

This is What Happens

The metal leaves move apart when the charged balloon is held near the metal loop.

Science Behind It

The instrument made is called an electroscope, or charge detector. The metal strips move when the loop is placed in an electrically charged area. Electrons are rubbed off of your hair and onto the balloon, charging the surface of the balloon with a negative charge. Holding the negatively charged balloon near the metal loop causes the negative electrons on the surface of the metal to move away from the balloon because like charges repel, that is, move away from each other. Electrons move down the metal wire and build up on the aluminum strips. Since like charges repel each other, the two negatively charged strips move apart.

22. Wrapping Reflection

In good old days, all wrapping paper used to be made of paper. Nowadays, you can get wrapping material that is made of very thin plastic. Some of this plastic material has a reflective silver covering. Perhaps you've received a birthday gift wrapped in this reflective material? If so, what did you do with the mirrored fabric after you ripped it off your present? Hopefully, you kept it around for this next experiment.

Material Required

- Silver-coated wrapping material;
- Glue stick;
- Index card;
- Pair of scissors.

Procedure

Cut out a card-sized rectangle of silver-coated wrapping material. Use a glue stick to coat an index card with a thin and uniform layer of adhesive. Place the sliver rectangle in the centre of the card. Gently press it from the centre of the card. Try to squeeze out any bubbles or regions where the material forms a bump or fold. Let it dry.

Examine the silver-coated card. Can you see your reflection in the card? Does this material make a good mirror?

While looking at your reflection, bend the card. Make the card into a convex mirror by having the centre bend towards you. Make the bend up and down. Make the bend sideways. What happens to the distortion as you change the angle of this bend?

Now bend the card into a concave mirror. Can you locate the viewing distance where your image flips upside down?

This is What Happens

A good mirror has a reflective surface that's both silver-coated and extremely flat. Glass that is coated with silver paint makes a great mirror. The wrapping paper, however, wasn't so perfect. Although the material was coated in silver, it wasn't very flat. Slight bumps in its surface produced major distortions that spoiled its reflecting possibilities.

23. Magical Wand

"Come to me."

"Go that way. I command you."

"Turn in a circle."

Have you ever wished for a magic wand? You know the type. You can use it to command the movement of all sorts of objects. Well, in this activity, you'll have a chance to build this type of wand. The "magic," however, is hidden in the invisible field of electric charges.

Material Required

- Table tennis ball;
- Plastic pen;
- Wool.

Procedure

Set the table tennis ball on a flat surface. Steady it so that it doesn't move.

Stroke the plastic pen with the wool. After stroking the pen, move it close to the table tennis ball. What happens?

Try moving the pen so that the ball moves with a continuous motion. Can you do it?

This is What Happens

As the pen was stroked with the wool, negative charges were transferred between the materials. These charges left the wool and accumulated on the pen. The pen became negatively charged.

When the negative pen was placed near the table tennis ball, its electric field affected the charges of the ball. The nearby negative charges were repelled by the pen and moved away. This left a relatively positive side of the ball. This positive side and the negative pen attracted. Once inertia and friction were overcome, the ball began moving.

24. Touch Me Not

Can you move a piece of paper in a glass with out touching it? Impossible will be your answer, but by performing this easy experiment you can do it. This experiment also demonstrates the force of attraction between charged particles.

Material Required

- Modeling clay;
- Push tack;
- Tissue paper;
- Scissors;
- Clear plastic glass;
- Balloon, small enough to hold in your hand;
- Ruler.

Procedure

Roll a dime-sized piece of clay into a ball and press it onto a table. Push the end of the push tack into the clay, leaving the point sticking up. Cut a 1 in. (2.5 cm) square from the tissue paper. Fold the paper in half to form a tent. Balance the paper tent on top of the pin point. Position the plastic glass over the pin and paper tent.

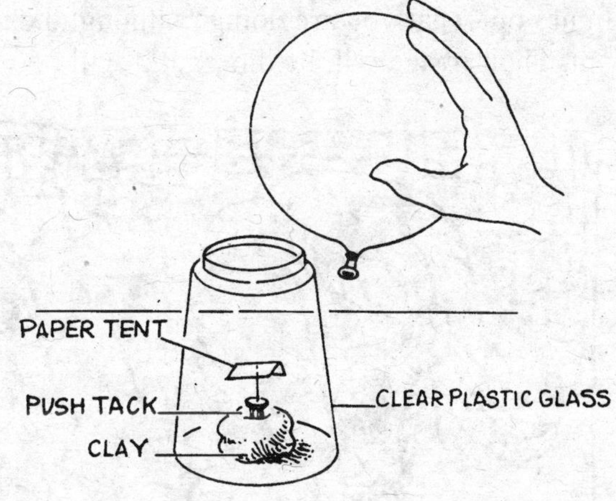

Inflate the balloon to a size that is easily held in your hand. Charge the balloon by rubbing it on your hair about 10 times. Your hair must be clean, dry, and oil-free. Hold the charged balloon near, but not touching, the plastic glass. Watch the paper tent for any movement.

This is What Happens

The paper tent turns and falls off of the tip of the tack.

Science Behind It

The paper tent, balloon, and hair are all examples of matter, and all matter is made of atoms. Atoms have positively charged protons in their centre (the nucleus) and negatively charged electrons spinning around the nucleus. The balloon becomes negatively charged on the side that is rubbed on the hair because the electrons are rubbed off the hair and collect on the balloon. The negatively charged balloon attracts the positive part of the paper tent. This attraction is strong enough to pull the paper off the tack.

25. One Way Mirror

This may sound strange to you; you may have noticed that in some shops a mirror is placed behind that check-out counter? There's no accounting for the manager's sense of decor, unless of course she's on the lookout for shoplifters.

Material Required

- Silver packaging wrap.

Procedure

Turn up the lights in the room you're in. Stretch the reflective sheet of silver wrap so that it is taut. Hold it at arm's l ength and examine your reflection. Weird.

Now bring the material close to your face so that it touches your nose. Can you see any objects in the room?

Call for a friend's attention. While holding up the material, make faces. Have your friend try to figure out what you are doing. Although he or she won't be able to see in, you'll be able to see out. Their image will be dim, but it will be visible!

This is What Happens

This wrapping material makes a great one-way mirror. This property is produced by a thin silvery coating. Like the one-way mirrored glass (used in stores), this material had a covering that wasn't thick enough to block all light. It allowed some light through.

To increase the one-way experience, objects on one side of the glass are brightly lit while the objects on the other side have less illumination.

26. Suspended Figures

Electrical charges flow easily through metal. Yet we haven't used metal in any of the experiments. Why? Because the problem with using metal is that—unlike materials such as plastic, paper, and nylon—metal is not lightweight. It's heavy. Metal's mass can easily "mask" the attraction of electrostatic forces. But for those of you who still want a metal experiment, here's this stand-up activity.

Material Required

- A sheet of light-gauge aluminum foil;
- Pair of scissors;
- Clear plastic food bowl with cover;
- Wool.

Procedure

Use your scissors to cut out several human-like figures from a sheet of light-gauge aluminum foil. The figures from head to toe should be slightly shorter than the depth of the plastic food bowl (from bottom to cover).

Place the figures in the bowl. Cover the bowl. Use the wool to vigorously stroke the bowl's cover. What happens?

This is What Happens

As the plastic cover was stroked, it acquired a negative charge. This charge induced a positive charge in the near ends of the aluminum figures.

Since the heads were the lighter end of the cutouts, they probably were the first part that moved upwards. Since the charge wasn't great enough to overcome all of the figures' weight, they did not fly up and cling to the cover. Instead, the aluminum cutouts remained suspended in air midway between the top and bottom of the bowl.

27. Flying Attraction

The idea behind this experiment is to demonstrate the effect of static electricity.

Material Required

- Comb;
- Aluminum foil;
- Scissors.

Procedure

Cut ten tiny pieces of aluminum foil and lay them on a table. Quickly move the comb through your hair. Your hair must be clean, dry, and oil-free. Hold the teeth of the comb above the foil pieces. Do not touch the aluminum.

This is What Happens

The aluminum foil pieces move towards the comb. The metal actually moves through the air to reach the comb.

Science Behind It

Aluminum foil is made of atoms, which are composed of positive parts called protons and negative parts called electrons. The protons are in the centre of the atom (the nucleus) and the electrons spin around the outside of the nucleus. The comb rubs electrons off your hair and becomes negatively charged. As the comb approaches the metal pieces, the negative electrons in the metal move away from the comb, leaving more positive charges on the surface of the metal. Like charges repel each other and unlike charges attract. The attraction between the negatively charged comb and the positive area on the metal is strong enough to overcome the downward pull of gravity, and the metal pieces move through the air to stick to the comb.

28. Compact Disk Copier

Walk by the large glass window of a store. Look into the display and most likely you'll see someone familiar staring right back out. Although glass is transparent, it has some reflective qualities. In this experiment, you'll use both properties of glass to construct a great copying machine.

Material Required

- Clear plastic CD case;
- Pen;
- Blank sheet of paper;
- Lamp (if available).

Procedure

Place a blank sheet of paper alongside an image you wish to trace. Remove the cardboard insert from a clear CD case. Open and position the case so that the clear plastic section is aligned along the border of the two papers.

Look into the plastic. Do you see a ghost-like reflection? You can increase the intensity of this reflected image by shining a bright light on the original copy while dimming the light over the blank copy sheet. By following the image outline, you can trace a copy of the original onto the blank paper. How is this copy similar to the original? How is it different?

This is What Happens

The clear plastic case has some reflective qualities. Light from the original drawing travelled to the plastic case. Some of it was reflected off the surface and back to your eyes.

When you looked through the plastic case, you were able to "project" this reflected image onto the blank sheet of paper. The drawing that you created was right/left reversed, since it was a copy of a reflected mirror image.

29. Humidity and Static Electricity

Sometimes, static electricity experiments work great. Other times, they seem to fizzle out. Here's an experiment that will help you explore how your surroundings affect static charge.

Material Required

- Balloon;
- Bathroom.

Procedure

Find a bathroom with a shower or bathtub. Make sure that no one has bathed in the room for several hours. The room's air should be dry.

Inflate a balloon. Charge the balloon by rubbing it on your hair. What charge does the balloon acquire? Place it against the wall. Observe how it clings to the surface.

Turn on the water in the shower or bathtub. Let the room fill with humid air. Now try charging the balloon again. Place it against the wall. What happens this time?

This is What Happens

Electric charges cannot flow well through dry air. As the balloon was rubbed against the hair, charges moved onto the balloon. However, the dry air insulated the balloon and prevented the charges from "leaking" into the surrounding air.

Electric charges flow much better in humid air. When the air was humid, charges leaked from the balloon. Since the balloon lost its charge buildup, the clinging effect (if any) was not very great.

30. Snapping Sound

Do static charges produce sound? This is what we are going to find out.

Material Required

- Large paper clip;
- Piece of wool;
- A scarf, coat, or sweater made of 100% wool will work;
- Clear plastic sheet;
- Scissors;
- Modeling clay.

Procedure

Cut a plastic strip about 1 in. x 8 in. (2.5 cm x 20 cm). Use the clay to stand the paper clip upright on a table. Wrap the wool around the plastic strip and quickly pull the plastic through the cloth. Do this quickly at least three times. Immediately hold the plastic near the top of the paper clip.

This is What Happens

A snapping sound can be heard.

Science Behind It

Electrons are rubbed off the wool and onto the plastic. The electrons clump together until the addition of their energy is great enough to move them across the span of air between the plastic and the metal clip. The movement of the electrons through the air produces sound waves, resulting in the snapping sound heard.

31. Mysterious Mirror

Smoke and mirrors. For centuries, this phrase was the magician's hallmark. It was believed that most stage illusions were accomplished with the aid of these tools. One of the more popular tricks of the 1800s was making your assistant disappear and reappear at will. Although we won't tell you how it was done, you might figure it out after performing this experiment.

Material Required

- Clear drinking glass;
- Plastic CD case;
- Matches;
- Plate;
- Candle.

Procedure

Remove the cardboard insert from a clear CD case. Open and position the case so that the clear plastic sections form a stable right angle. Place a drinking glass on the near side of the clear plastic. Place a candle on a plate and position them on the opposite side of the CD case.

Have an adult light the candle. From the drinking glass side, look into the plastic and find the reflection of the glass. Move the glass so that its reflection aligns with the lit candle. Dim the lights in the room. When the glass is in the correct position, the reflection and transmitted image will combine to make it appear as if the candle is burning within the glass. Magic? No—it's science!

This is What Happens

The image that you see in the CD case is a mix of both reflected light and transmitted light. The reflected light comes from the drinking glass. The transmitted light comes from the candle. Since both images overlap, your brain assumes that they came from the same location and places the burning candle in the glass.

32. Sparking Doorknob

In this experiment, you will find something shocking and snapping.

Material Required

- Wool rug;
- Metal doorknob.

Procedure

Find a room that has a rug and a metal doorknob. Turn off all of the lights and pull down the window shades. Put on your shoes and walk across the rug to the door.

Slowly bring your index finger towards the doorknob. What did you hear? What did you feel? What did you see?

This is What Happens

As your shoes rub against the rug, negative charges leave it and enter your body. Although you don't feel it, your body takes on a net negative charge.

In nature, things like to be stable. Electric charges will move to create a stable, balanced condition.

As your hand nears the doorknob, negative charges concentrate at your fingertips. Just before you touch the knob, the charges have enough energy to "jump" across the gap. This jump produces a tiny spark. The spark heats up the air and produces a "snap."

33. Balloon Attraction and Repulsion

The purpose of this neat experiment is to demonstrate attracting and repelling forces between objects due to their electrical charges.

Material Required

- 2 round balloons, 9 in. (23 cm);
- Masking tape;
- String, 2 yd. (2 m);
- Clean, dry, oil-free hair;
- Marking pen.

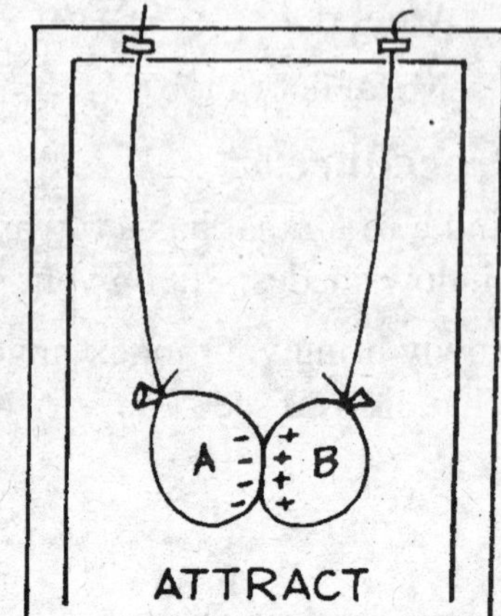

Procedure

Inflate both balloons and tie their ends. Use the marking pen to label one balloon A and another B.

Cut the string in half and attach one piece to the end of each balloon. Tape the free ends of the strings to the top of a door frame so that the balloons hang about 8 in. (20 cm) apart.

Stroke balloon A on your hair about 10 times and gently release it. What happens?

Stroke one of the balloons on your hair about 10 times and hold it while an assistant rubs the second balloon on your hair. Gently release both balloons. Now what happens?

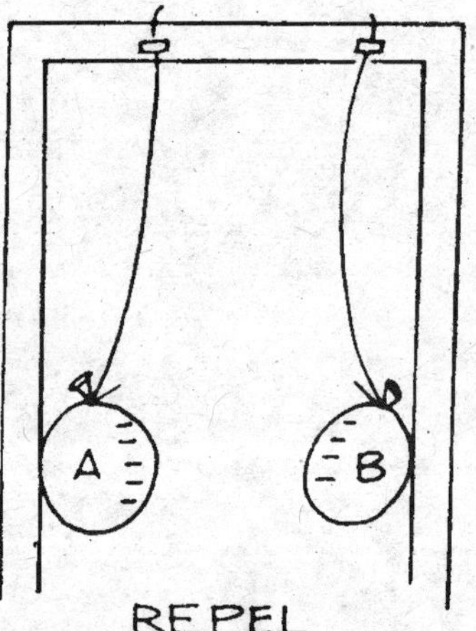

This is What Happens

The two balloons are attracted to each other when only one balloon is stroked against hair and repel each other when both balloons are rubbed against hair.

Science Behind It

Matter is made up of atoms, which have negatively charged electrons spinning around a positive nucleus. Electrons are rubbed off the hair and collect on balloon A; thus the balloon becomes negatively charged. Since like charges repel each other, these negative charges on balloon A repel the electrons in the atoms of balloon B, causing B's surface to be more positively charged. Each balloon now has a different charge, so they are attracted to each other.

Rubbing both balloons on the hair results in a build-up of negatively charged electrons on their surfaces. The balloons repel each other because they have the same charge. There are enough repulsive and attractive forces between the balloons to cause them to move without being touched.

34. Reappearing Coin

For our next trick, we are going to make one million rupees appear from nowhere! Sorry, we are mistaken. Due to tax laws, we won't be able to work with such a large sum of money. We will, however, make a hidden coin magically reappear.

Material Required

- Opaque soup bowl or mug;
- Water-filled cup;
- Coin.

Procedure

Place a coin in the bottom of an opaque soup bowl or mug. Position yourself off to the side. Slowly lower your viewing angle until the coin can no longer be seen. What happened to the coin? Did it disappear?

Suppose light rays didn't travel in straight lines. How might a curved ray affect your observations? While keeping your viewing angle, slowly add water from the cup into the bowl.

As the level of water rises, you may be surprised to observe your "lost" coin. What caused the coin to reappear?

This is What Happens

Although light rays travel straight paths, these paths change direction as light moves from one substance to another. The border (scientists often call this an interface) between the water's surface and the surrounding air causes light rays to bend. So when you look at the water's surface, you don't see objects that are in a direct line-of-sight. Instead, you see things that are submerged and out of a direct line-of-sight.

35. Sparking Clothes

Have you ever tried to separate clothes others that were just taken out of a drier? If so, you most likely encountered static cling. As the clothes were tossed in the drier, the materials rubbed against each other and transferred charges. Clothes that became oppositely charged stuck together.

Material Required

- Silk shirt;
- Wool sweater;
- Mirror.

Procedure

Put on a silk shirt. Pull a wool sweater over the shirt.

Find a room that has a mirror. Turn off all of the lights and pull down the window shades. The darker the room, the better the effect.

Stand a few feet in front of the mirror. Slowly roll up you sweater. What do you observe?

This is What Happens

As the wool sweater rubbed against the silk shirt, charges were transferred. The shirt became negative in charge. The sweater lost electrons and became positive. The negative shirt and positive sweater attracted each other.

As you lifted the sweater, the charged materials separated. As they were pulled apart, visible sparks jumped the tiny gaps. These sparks produced a crackling sound.

36. Tapes that Don't Stick

Cellophane tape is allway sticky. But in this experiment, they seem to behave odd. In this experiment, we will understand more about the effect of static electricity.

Material Required

- Cellophane tape.

Procedure

Press two pieces of tape onto a table, leaving a small piece hanging over the edge. Hold the ends of the tape and quickly pull both pieces off of the table. Bring the two pieces near each other, but not touching.

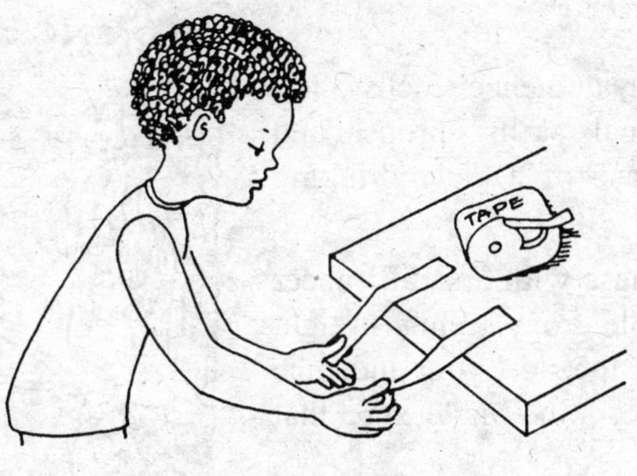

This is What Happens

The pieces of tape move away from each other.

Science Behind It

All materials are made up of atoms, which have positive and negative parts. When there is a gain or loss of charges, the object is said to have static electricity. Pulling the tape pieces from the table causes them to pick up electrons from the atoms in the table. Both pieces of tape are negatively charged. Materials with like charges repel (push away from) each other.

37. More Coin Magic

If you have you liked the last trick. Now you want another magic trick to play on your little brother or sister. How about a disappearing act? No, we won't make him or her disappear. Instead, you'll focus your art of illusion on a less pestering subject--another coin.

Material Required

- Two clear drinking glasses;
- Saucer;
- Water;
- Coin.

Procedure

Be theatrical. Have your audience focus their attention on your magical skills, props, and presentation. Place the base of a clear drinking glass over a coin.

Cover the mouth of the glass with a small saucer (the saucer prevents people from seeing straight down into the glass). From the side, your audience members will observe the coin. Make sure that everyone can see the coin.

Now tilt back the saucer and fill the glass with water. Again, cover up the mouth of the glass with the saucer. Can the audience still see the coin? What happened to it?

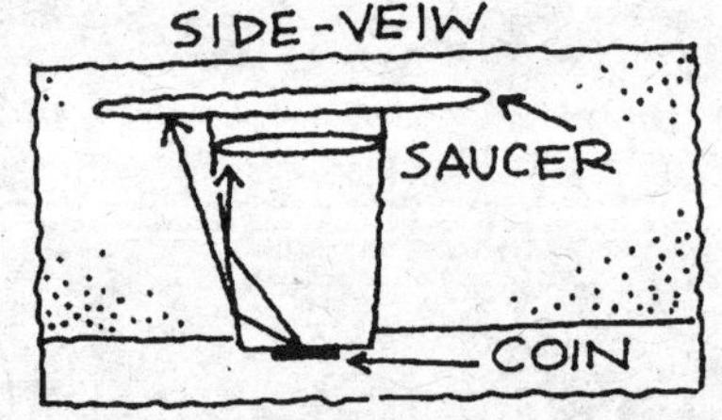

This is What Happens

It's those bending light rays again. The image of the coin travels along straight-line paths. When the glass is empty (actually it's filled with air), there is little distortion and the rays travel a mostly straight path from the coin to your eyes. The water changes things.

Now, the light rays first travel in water. When they strike the side of the glass, some of them are bent and travel up the side of the glass. The light rays that do escape from the glass are bent so much that they travel almost straight up and are blocked by the part of the saucer that extends past the mouth of the glass. Since the rays carrying the coin's image can't get to your eyes, the object has "disappeared."

38. Rattling Radio

S…0...S. As the Titanic sank, its radio officer sent out this distress call. Each click of the Morse code's key temporarily closed an electric circuit. This circuit produced a spark, which created an invisible energy wave that travelled out from the sinking ship. These waves were detected by the antennae of other boats. The antennae's signal was carried by wire to a receiver. There, the invisible waves were changed into an audible click.

Here's an experiment that illustrates how sparks were used to send Morse code messages. All you'll need is a carpet, doorknob, and a radio.

Material Required

- Carpet;
- Metal doorknob;
- Radio.

Procedure

Switch on a radio. Tune the radio to a frequency where no signal is detected. With the volume turned up, the radio should broadcast a low level of static.

Walk across a wool carpet wearing a pair of shoes. Approach a doorknob. While listening to the radio, reach out and touch the knob. What do you hear on the radio?

This is What Happens

Sparks create a form of energy called an electromagnetic wave. This wave travels outward through space. The antenna of a radio can detect this form of energy. The signals it captures are carried along wires to the radio's circuitry. There, the signals are turned into sound, which becomes amplified and broadcast through the radio's speakers.

39. Right Hand and Apple

Can you find the magnetic north pole by the help of apple or tennis ball? In this easy experiment, we will show you how you can do it.

Material Required

- Apple or tennis ball.

Procedure

Hold the apple in your right hand, and assume it is the earth, spinning on its axis. Wrap your fingers around the sides of the apple, with your thumb sticking up.

This is What Happens

Your fingers point in the same direction that the earth spins. Your thumb points in the same direction as does the earth's magnetic north pole.

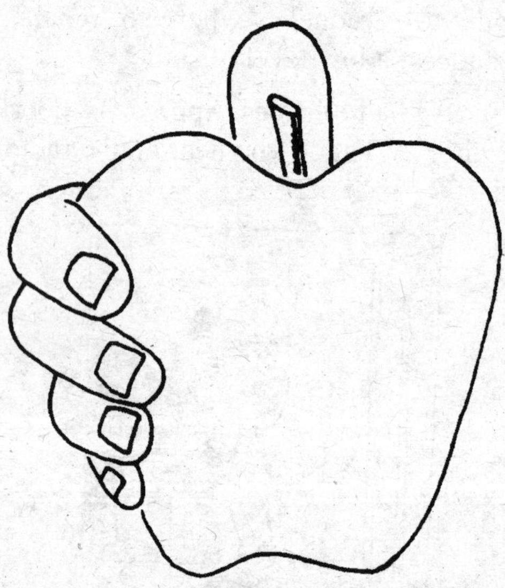

Science Behind It

The entire earth, including the liquid core, spins in a counterclockwise direction. Electrons are rubbed from this turning liquid, but they continue to move in the same direction as the earth. A magnetic field is always present around moving electrons. The direction of the magnetic field can always be determined by using your right hand. Point your fingers in the direction of the moving electrons, and your right thumb will indicate the direction of the magnetic north pole.

40. All Out

You might have seen some of the popular 3-D movies like "Chota Chetan." Wow! That guy jumped right out of the screen! How did they make those 3D effects? If you tried to remove your viewing glasses, you probably were disappointed. The movie looked out of focus. Actually, there were two movies which were projected onto the same screen. The eyeglasses contained polarizing lenses. These special lenses separated the overlapping movies and sent a different image to each eye. Your brain did the rest and produced the illusion of 3D.

Material Required

- Two pairs of polarizing sunglasses (not just any sunglasses—the lenses must be made of polarizing material).

Procedure

Close one eye. Use your open eye to look through a polarizing lens. What do you see? A slightly darkened world? Good.

Now hold up the second lens right behind the first. Look through both lenses. What do you see now?

Slowly rotate one of the lenses. What happens to your view? Can you rotate one lens so that no light passes through to your eyes?

This is What Happens

Make a "thumbs up" sign with your hand. Good. Now extend the index finger of this hand. Your index finger represents the direction that the light ray travels. Your thumb points to the direction at which this light wave vibrates.

Rotate your wrist. Notice how the ray still travels in the same direction. However, the vibrations project in all directions along this circular path. The thumb's side-to-side waves can be cut out using polarizing filters. A polarizing filter can limit the vibrations to mostly one plane. This cuts down on the transmitted light.

When two filters are used, you can do even more. When the filters are positioned so that one knocks out up-and-down vibrations while the other cuts out side-to-side vibrations, no light is transmitted.

41. Jumping Puffs

Puffed cereal grains are great materials or science experiments. Since they are light in weight, they do not require much force to move. In addition, the puffs easily transfer electric charges. Want to see? Just try this next experiment.

Material Required

- Puffed cereal grains;
- Balloon;
- A piece of wool or fur.

Procedure

Stuff about a dozen grains of puffed cereal into a balloon. Inflate the balloon.

Rub the balloon with a piece of wool or fur. If the material isn't available, you can rub the balloon against your hair.

Hold the hanging balloon by its knot. Observe the grains within the balloon. Are they stationary or moving? Touch the balloon with the fingertips of your other hand. How do the grains behave? If nothing happens, recharge the balloon by giving it twice as many strokes. Then, touch it again.

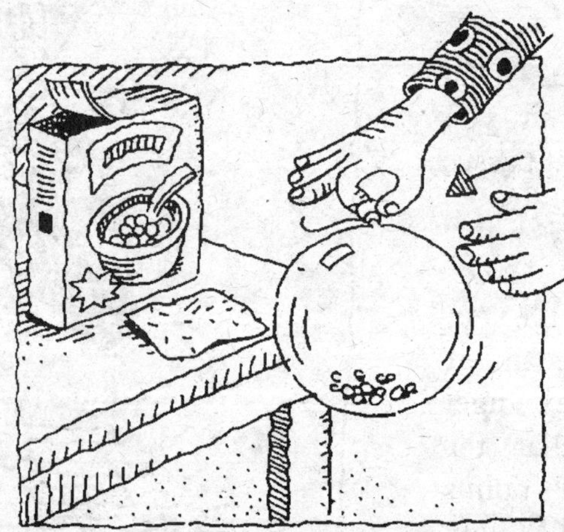

This is What Happens

As the balloon rubbed against the wool, it became negatively charged. Its negative field induced a positive charge in the nearby side of the puffed grains. This positive region was attracted to the balloon, causing the grains to cling to the balloon's negative skin.

When you touched the balloon with your fingertips, things changed. The balloon's negative charges drained out through your fingers. This created a positive region in the balloon. The charges in the grains could not shift fast enough. Instead, the positive grain surface and the positive balloon skin repelled each other. The grains jumped to another location.

42. Magnetic Paper Clip

The function of this experiment is to determine if the north end of a magnet always points to the earth's magnetic north pole.

Material Required

- Compass;
- Sewing thread;
- Small paper clip;
- Cellophane tape;
- Ruler;
- Magnet;
- Book.

Procedure

Cut a 12 in. (30 cm) piece of thread. Attach one end of the thread to the centre of the paper clip with a very small piece of tape. Tape the free end of the thread to the end of the ruler. Set the book on the edge of a table and insert the free end of the ruler between the book's pages so that the ruler hangs over the edge of the table. Place the paper clip on the magnet. Remove the clip from the magnet and allow it to swing freely. Observe the direction towards which the end of the paper clip points. Use the compass to determine what direction that is. Move the ruler to different positions and observe the direction the paper clip points each time.

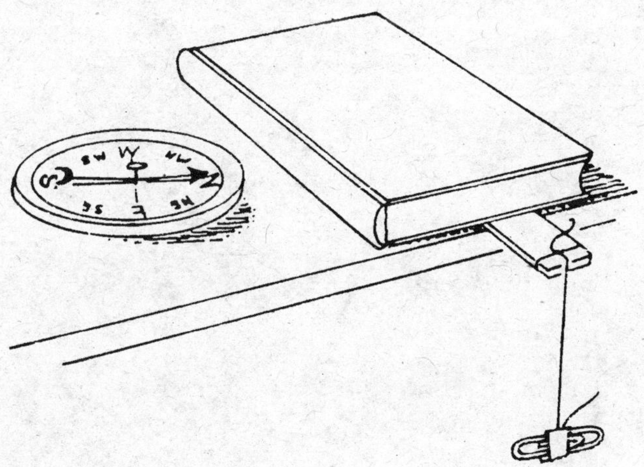

This is What Happens

One end of the paper clip points south and the other points north. Moving the ruler does not affect the direction that the paper clip points.

Science Behind It

The earth behaves as if a large bar magnet were inside it causing magnetic materials to be attracted to its opposite ends. The south end of this imaginary magnet produces the earth's north magnetic pole, and the north ends of all magnets are attracted to this pole. The north ends of magnets are really north-seeking poles. Placing the paper clip on the magnet causes the atoms inside the paper clip to line up in a north-to-south direction. One end of the paper clip will continue to point towards the earth's magnetic north as long as the atoms inside are lined up. Moving the position of the ruler or string will not affect the direction of the free-hanging magnetized paper clip.

Note: The paper clip will remain magnetized for only a short period. You may need to put the paper clip on the magnet to realign the electrons.

43. Bending Rays

Using Ray Marker which we have made in our earlier experiment, we will try to redirect the rays in this experiment.

Material Required

- Ray-making tool (from "Ray Marker");
- Hand lens;
- Clay;
- Flashlight;
- Clear drinking glass;
- Water;
- A drop of milk.

Procedure

Fill a glass three-fourths full with water. Set this water-filled glass on a flat surface.

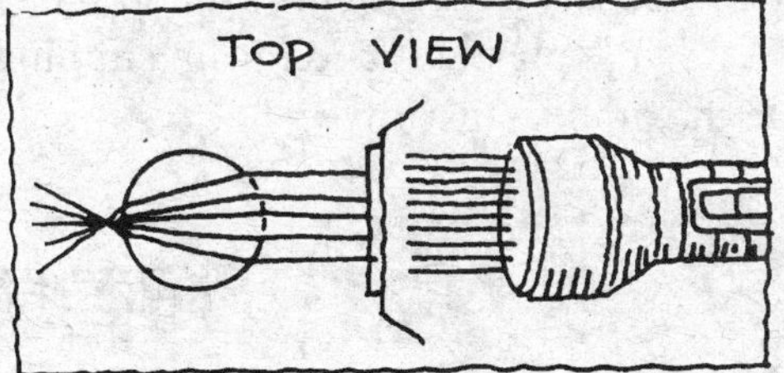

Position the ray-making tool several inches from the glass. Turn on the flashlight and aim its beam through the comb's teeth. The separate rays should illuminate their path along the tabletop.

Position the glass so that these rays strike it in the centre. You may have to raise the ray maker on a book to obtain the best beam angle. Observe the beams from above. What happens to rays as they strike the water-filled glass? How many times does their angle change? To help see the rays, make the water partly cloudy by adding a drop of milk.

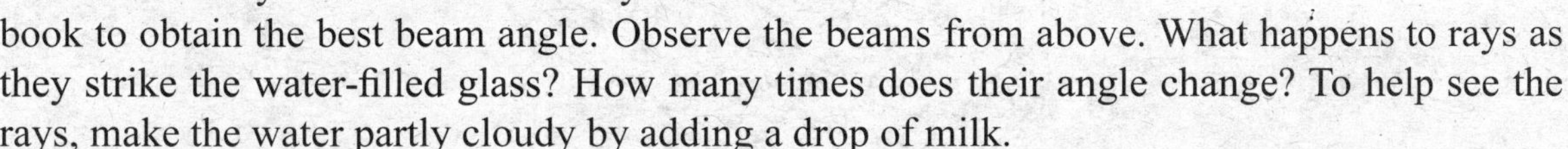

Replace the water-filled glass with a hand lens. Position the lens so that the separate beams can be observed striking the centre of the lens. What happens to these rays? Does the hand lens affect the rays in the same fashion as the water-filled glass?

This is What Happens

As the mostly parallel light beams struck the water-filled glass, their direction changed. The first change occurred at the point where they entered the vessel. The collision with both the glass container and water caused the light rays to bend together.

As they travelled through the water, the rays followed a straight path. When they exited the glass, their direction changed once more. Again, the rays wound up closer together. Several inches past the glass, the rays actually crossed over each other to form a bright spot. This special point is called the focus point. At the focus point, light rays are the most concentrated. Beyond the focus point, light rays spread out.

44. Showering Puffs

In this experiment, you will see static shower of the puffs.

Material Required

- Section of Lucite;
- Two thick books (or stacks of smaller ones);
- Puffed cereal or pieces of plastic foam.

Procedure

Place two thick books about 8 inches apart on a tabletop. The top of the books should reach several inches above the table.

Scatter puffed cereal grains or pieces of plastic foam in the gap between the books. Place a piece of Lucite across the gap so that it rests firmly on the books.

Rub a piece of wool or felt across the surface of the Lucite. What happens to the grains below? If you observed some movement, try rubbing your hand back-and-forth across the Lucite. To prevent the plastic section from slipping, you'll need to steady the Lucite with your other hand. Do the grains still move?

This is What Happens

As you stroked the plastic, it became charged through contact. This plastic's negative field affected the charge balance in the cereal grains. The grains' nearby negative charges migrated towards the farther side of the grain. This created a positive end of each grain that was attracted to the plastic. This attraction caused the grains to jump upwards and cling to the Lucite.

While the grains were attached to the Lucite, the balance changed. Electrons in the plastic moved onto the grains. Although it took them several movements to transfer, this flow eventually canceled the grains' positive surface. Without this opposite charge, the grains were into longer attracted to the plastic. They dropped off because of their weight.

45. Floating Needle

Can magnetic force move a floating object? We will try to find answer to this question by performing this experiment.

Material Required

- Glass bowl, 2 qt. (2 litre);
- Sewing needle;
- Sewing thread;
- Masking tape;
- Bar magnet.

Procedure

Fill the bowl three-quarters full with water. Cut two 12 in. (30 cm) pieces of thread. Tape both pieces of thread to one side of the bowl, about 1 in. (2.5 cm) apart. Stretch the thread across the bowl and lay the needle across both pieces of thread. Slowly lower the thread until the needle rests on the water's surface. Gently move the thread from under the needle. Move the magnet near, but never touching the floating needle.

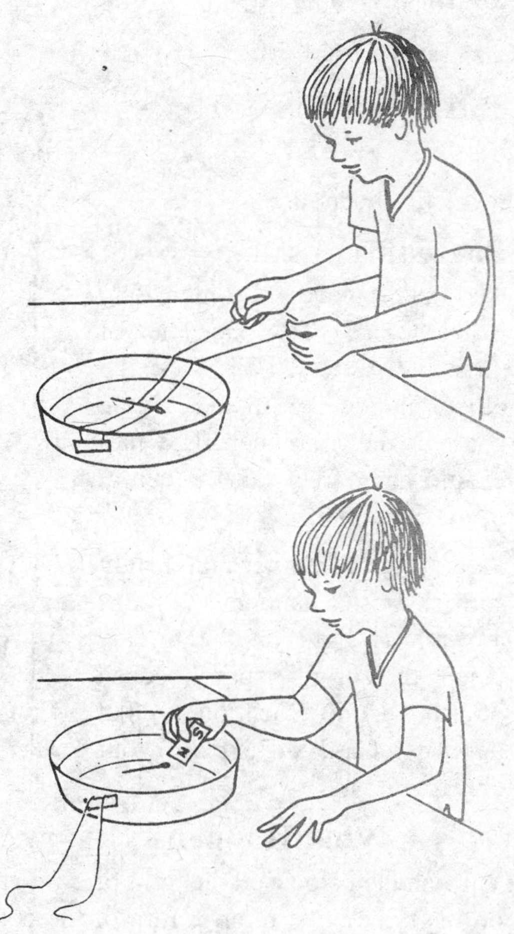

This is What Happens

The needle floats on the surface of the water and moves when the magnet moves.

Science Behind It

The surface of the water acts like a thin skin. This imaginary skin is due to the attraction of water molecules to each other. The needle is able to float and move across the surface of the water in response to the magnetic force of attraction by the magnet.

46. Light Heat

Close your eyes and look skyward. You can probably tell where the Sun is without looking. The warmth from its rays gives away the position of this nearby star.

Imagine what would happen if you were able to concentrate the Sun's rays. You'd produce a very bright spot. You'd also produce enough concentrated sunlight to burn through a string!

Material Required

- Plastic container;
- Black sewing thread;
- Tape;
- Hand lens;
- Nail;
- Pair of scissors.

Procedure

Cut out a piece of 6-inch-long black thread. Tie one end of thread to a metal nail. Tape the other end of the thread to the neck of the container. The nail should hang freely in the centre of the container.

Place the container in bright sunlight. Use a hand lens to focus a beam of light on the thread. Keep the bright spot focused on the same location. What happens? Can you explain your observations?

This is What Happens

The hand lens focused the light rays into a very bright and hot spot. When this spot was focused on the thread, the temperature of the cotton material increased. Soon, the thread became hot enough to smoke. When enough of the fibres had burned, the thread broke and the nail fell to the bottom of the container.

47. Antigravity Nuts

As the magician chanted the magic words, the rope responded. Slowly, it appeared above the basket's rim. Up went the rope. Higher and higher it climbed, until its entire length stretched before the audience.

Material Required

- Plastic foam "peanuts";
- 1-foot-long thread;
- Tape;
- A piece of wool or fur;
- Balloon.

Procedure

Break off a segment of thread about a foot in length. Tie one end of the thread to the peanut. Tape the other end of the thread to the edge of a table. The peanut should hang down from the edge.

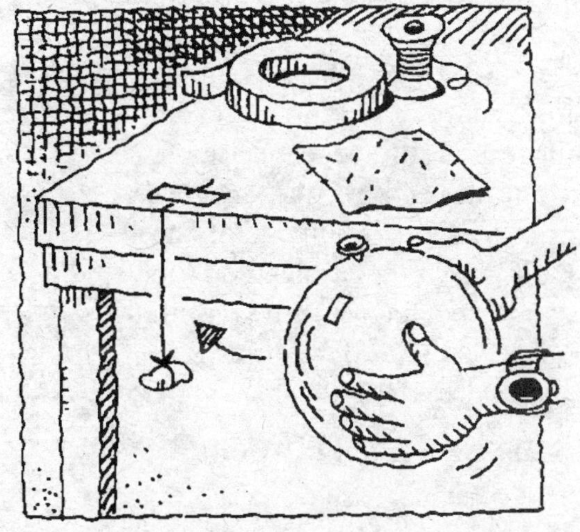

Rub the balloon with a piece of wool, fur, or your head. When the balloon is charged, slowly bring it closer to the peanut.

The peanut will be attracted to the balloon. As it moves closer to the balloon, slowly raise the balloon. Make sure that the peanut and balloon don't touch, or else an unwanted transfer of charges may occur. Keep raising the balloon until the peanut and thread extend perfectly upright. Keep raising the balloon. How far can it be raised and still support the peanut?

This is What Happens

The peanut starts out with its charge distributed evenly over its length. The negatively charged balloon induces a positive peanut region. This positive region and the negative balloon attract.

Since the peanut and thread are relatively light, the force of attraction can produce an interesting effect. The peanut and thread overcome gravity and lift off the table.

48. Hovering Airplane

In this experiment, we will use the magnetic force to suspend a paper airplane.

Material Required

- Steel straight pin;
- Sewing thread, 12 in. (30 cm);
- Tissue paper;
- Bar magnet;
- Scissors.

Procedure

Cut a small wing about 1 in. (2.5 cm) long from the paper. Insert the pin through the centre of the paper wing to make an "airplane." Tie the thread to the head of the pin. Place the magnet on the edge of a table with the end of the magnet extending over the edge of the table. Place the airplane on the end of the magnet. Slowly pull on the string until the airplane is suspended in the air.

This is What Happens

The airplane remains airborne as long as it stays close to the magnet.

Science Behind It

The strength of attraction between two magnets depends on how orderly the magnetic domains (clusters of atoms that behave like tiny atoms) are in the magnets. The atoms in the pin are randomly arranged before the pin touches the magnet. The number of atoms that arrange themselves into clusters (domains) and line up in the pin when it is placed on the magnet depends on the strength of the magnet.

The pin and magnet both have magnetic properties. They pull on each other with enough force to overcome the downward pull of gravity, which allows the airplane to remain suspended.

49. Pencil Pains

You don't have to be in the desert to see a mirage. Just look down a long road on a hot summer day. Most likely you'll see what looks like water or puddles on the road.

As you approach this slick strip of road, however, it dries up. What happened? You just experienced a mirage. This mirage was caused by the bending of light rays. What appeared as water was actually blue sky. The heated atmosphere produced a type of lens that bent the light rays. When you looked down the road, the wavy sky appeared on the ground instead of up above.

Material Required

- Pencil;
- Clear tall drinking glass;
- Water.

Procedure

Fill a drinking glass halfway with water. Place a pencil in the glass. Describe the appearance of the pencil. Is it straight or does it appeal to bend or break at the spot where it enters the water?

Slowly fill the glass with water. What happens to the "break" as the water level rises?

This is What Happens

The glass and water act as a light-bending lens. Your eyes detect the bent light rays but assume that they travelled on a straight path. Your brain "projects" where a straight path had come from. Mistakenly, it places the submerged pencil in a position where it's not.

In contrast, there is no confusion above the water's surface. This light travelled without bending. Your brain correctly projects where this part of the pencil was. The "break" arises from the two different projections.

50. Making a Versorium

A versorium is a device that is used to detect a static charge. Its name means "thing used for turning." It was given this name by its inventor about 400 years ago. Although times have changed, the principles that affected this tool haven't.

Material Required

- Metal paper clip;
- Paper;
- A piece of wool or felt;
- A pair of scissors;
- Plastic comb or pen.

Magic Words: Doing the experiment in a transparent cup will prevent breezes from upsetting the versorium's delicate balance.

Procedure

Unbend the larger arch of a paper clip. Position the smaller arch as a flat base. The unbent section should project straight up.

Draw the pattern shown below on a sheet of paper and use a pair of scissors to cut it out.

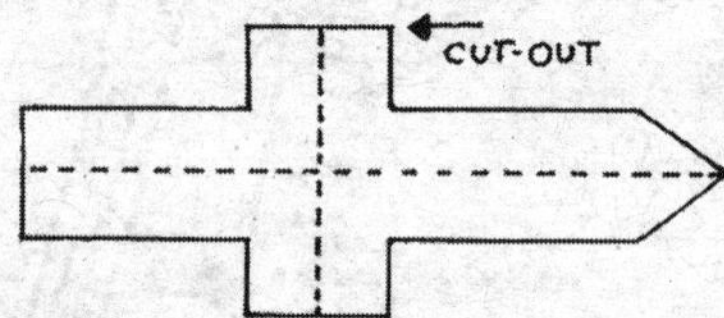

Put a slight downward crease along the dotted lines. Where they meet is the centre of balance. Gently balance this pointer onto the paper clip.

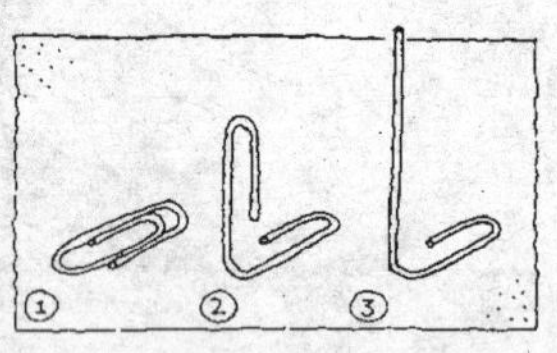

Charge up a plastic comb or pen with a piece of wool or felt. Bring the pen close to the versorium. What do you observe? Can you get the pointer to spin in an entire circle?

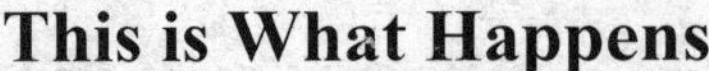

This is What Happens

The charged comb induced a positive region in the folded paper. This positive region and the negative comb attracted each other. The force was great enough to spin the pointer in any direction.

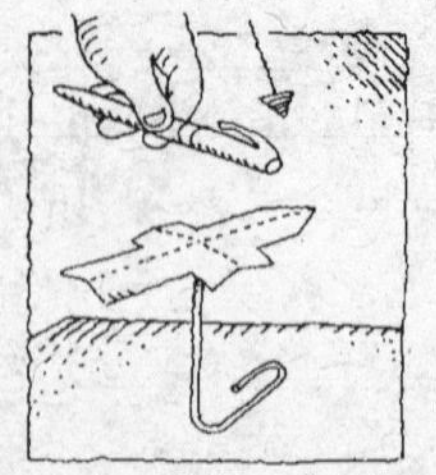

51. Magnetic Muscle

How can you find out the strength of a magnet? In the simple experiment, you will learn more about the strength of a magnetic force field.

Material Required

- Box of small paper clips;
- Bar magnets, several different sizes;
- Masking tape.

Procedure

Tape the magnet to a table with part of the magnet extending over the edge of the table. Bend open the end of a paper clip and touch it to the bottom of the part of the magnet that extends over the table's edge. Add paper clips one at a time to the open clip until the clips pull loose from the magnet and fall. Repeat this experiment with magnets of different sizes.

This is What Happens

The open paper clip hangs freely under the magnet. It continues to hang attached to the magnet as additional paper clips are added. The number of clips needed to cause the clips to fall will vary with different magnets.

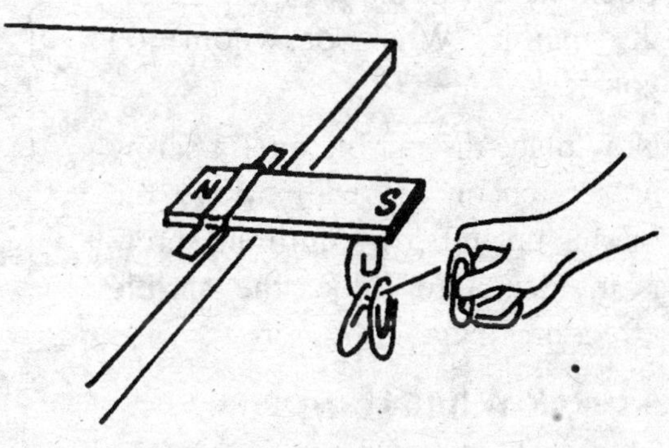

Science Behind It

Magnetic materials like steel contain clusters of atoms that behave like tiny magnets. These atomic clusters are called magnetic domains. Holding the paper clips near a magnet causes the magnetic domains inside to line up. The strength of the magnet increases as the number of domains pointing in the same direction increases. A weak magnet has a weak magnetic field around it, so its effect on magnetic materials such as paper clips is small. The number of paper clips that your magnet is able to support depends on its magnetic strength.

52. Magical Marble

Your eye contains a part called a lens. Like the lens of a camera, this eye part focuses light. But unlike the camera's glass lens, yours is, made of living muscle. Like other types of muscle, it can stretch and contract. By changing its shape, the lens can focus distant or close-up objects onto the light-detecting part of the eye called the retina without the action of the lens, our visible world would be a blur.

Material Required

- White wall;
- Clear glass marble.

Procedure

Stand next to a white wall. Make sure your body is well-illuminated. Hold out a clear glass marble several inches from the wall's surface. Look closely at the wall just below the marble. What (or whom) do you see?

Now place the marble over a screw or scratch on the wall. Move the marble up and down until the screw is in focus. How does the marble affect the image of the screw?

This is What Happens

The marble acts like a lens. Light rays that are reflected off your body enter the marble. They exit the marble and are cast onto the wall. If the marble is placed at the correct distance, a finely focused (and very reduced) image of your body can be seen on the wall screen.

At the same time your body shrinkage was occurring, the marble was magnifying wall objects. The screws and scratches appeared larger because the marble had spread out their reflected light rays.

53. Rising Ruler

Although gravity can be an incredible force, it requires huge masses to readily observe its effects. Not so with electricity. In fact, a minor charge can easily be observed using a tool that is often associated with uncovering an object's length.

Material Required

- Wooden ruler;
- Balloon.

Procedure

Position the wooden ruler so that it balances on the edge of a table. At this balance point, the ruler remains steady with one end slightly lifted from the tabletop.

Charge up the balloon by stroking it through your hair or with a piece of wool. Slowly approach the raised end of the ruler with the charged balloon. What happens? How does the electrostatic force work against the force of gravity?

This is What Happens

When the ruler was balanced, it remained still. As the charged balloon was brought over one end of the ruler, it upset the wood's balance of charges. The negatively-charged balloon induced a positive charge in the near end of the ruler. This positive charge and the negative balloon attracted each other. The force of attraction was large enough to offset the balance of mass and cause the end of the ruler to rise up.

54. Magnetic Patterns

The purpose of this easy experiment is to demonstrate the pattern of magnetic force fields around magnets of different shapes.

Material Required

- Magnets, a variety—round, bar, U-shaped;
- Iron filings;
- Notebook paper;
- Paper cup.

Procedure

Pour the iron filings into the paper cup. Place the magnets on a table. Cover the magnets with a sheet of paper. Sprinkle a thin layer of iron filings on the paper over the magnets. Observe the iron filing patterns.

This is What Happens

The iron filings form a pattern of lines around the magnets. The long magnet has a buildup of filings around both ends.

Science Behind It

A magnetic field is the area around a magnet in which the force of the magnet affects the movement of metal objects. The iron filings are pulled towards the magnets when they enter the magnetic field. The magnetic force increases as the filings near the magnet. The force field has equal strength around the round magnet, but the force fields at the ends of rectangular magnets are always stronger than the force fields in the middle of the magnets.

55. Pinhole Viewer

Take a look around you. Everything appears right-side up. Although this makes sense, it's not how your eye sees it. Images that are cast upon your eye's light-sensitive screen (called the retina) are upside down. It takes a learned brain trick to flip these images right-side up so that they match the look of the real world.

Material Required

- Plastic cup;
- Wax paper;
- Rubber band;
- Pushpin;
- Pair of scissors.

Procedure

Use the pushpin to punch a hole in the centre of the cup's bottom. Cut out a piece of wax paper slightly larger than the cup's mouth. Stretch this paper across the mouth of the cup. Use a rubber band to secure the paper.

Darken the room. Aim the pinhole at a bright window or lamp. Look at the wax paper. What do you see? How does the image compare to the actual scene?

This is What Happens

You've built a pinhole viewer. Light that came through the window entered into the hole of your viewer. After entering through the hole, the light rays continued on a straight path. This caused the upper light rays to strike the bottom part of the wax paper screen. Likewise, the lower light rays struck the upper part of the screen. The image that these rays produced was flipped upside down.

On the opposite page is a ray diagram that shows what happened when light from an object passed through the pinhole and was projected onto the viewing screen.

56. Repelling Nuts

As you've learned, like charges repel while unlike charges attract. You've also observed how charges can be transferred from one object to another. The following experiment uses both of these concepts. After building it, figure out how this tool might be used to measure charges.

Material Required

- Metal clothes hanger;
- 12-inch-long thread;
- Two plastic foam "peanuts";
- Balloon.

Procedure

Hold the hanger's hook with one hand. Firmly grasp the long side of the hanger with the other. Slowly and steadily stretch out the shape of the hanger until it looks like a stretched-out diamond.

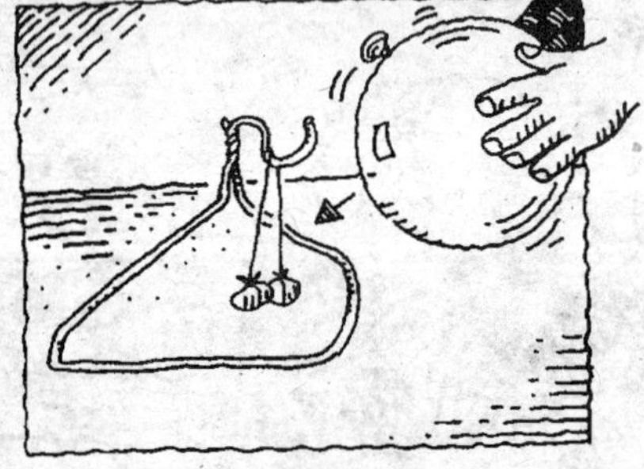

Bend up one half of the diamond. The fold you produce will form a stand. Give the hook a quarter turn so that it falls back towards the centre of the stand.

Tie a foam peanut to both ends of a 12-inch thread. Drape the thread across the upturned hook of the hanger. Position the thread so that the peanut hangs at the same level.

Inflate a balloon. Charge the balloon by rubbing it with a piece of wool or fur. Bring the balloon to the peanuts. How do the peanuts react? Touch the charged balloon to the peanuts. What happens now?

This is What Happens

The lightweight peanuts were easily influenced by electric charges. When the balloon was brought nearby, it caused the charges within the peanuts to separate. The region of the peanuts nearest the balloon became positive and was attracted to the negative balloon.

When the peanuts were touched by the charged balloons, electrons flowed. These electrons caused both peanuts to become negatively charged. Since they had the same charges, the peanuts repelled each other. The repulsion produced enough force for the peanuts to separate and rise. The distance that they separated depended on how much charge they acquired. This distance can be used as a method for measuring charges.

57. Magnetic Shake-up

Does shaking has any effect on a magnet? This is what we are going to find out.

Material Required

- Iron filings;
- Compass;
- Magnet;
- Drinking straw;
- Modeling clay.

Procedure

Fill the straw three-quarters full with iron filings. Use the clay to seal both ends of the straw. Lay the iron-filled straw on the magnet for one minute. Being careful not to shake the straw, pick it up and move it near the compass. Observe any movement of the compass needle. Shake the straw several times, and again hold the straw near the compass. Observe any movement of the compass needle.

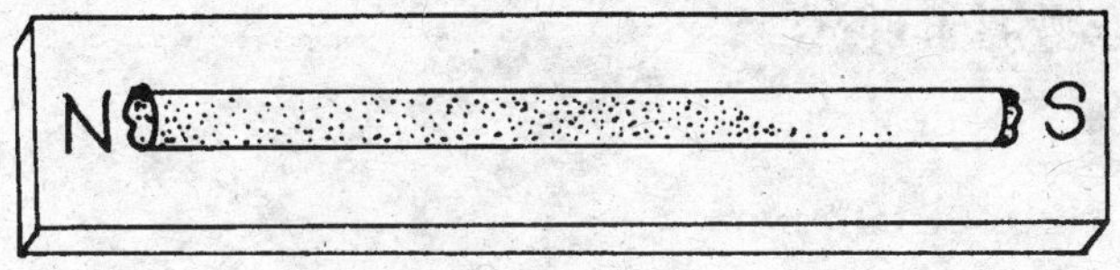

This is What Happens

The compass needle is attracted to the straw before the straw is shaken, but there is no movement of the compass needle after the straw is shaken.

Science Behind It

Magnetic materials have magnetic domain; (clusters of atoms that act like tiny magnets) that point in the same direction. The iron filings become magnetized when brought near the magnet because the magnetic domains in the iron filings line up. The compass needle is attracted to the magnetized, iron-filled straw. Shaking the straw rear ranges the pieces of iron, causing the magnetic domains in the iron to be randomly arranged, and thus the iron filings lose their magnetic properties.

58. Making Projector

The projection room at the back of a movie theatre is a cool place, literally. That's because the room has an incredible network of cold air pipes. This frigid air, however, isn't meant to keep the projectionist comfortable. It is sent straight to the movie-making machines. There, it cools the projectors and movie film. Without this continual blast of frigid air, the projector and film would be ruined by the intense heat given off by the white-hot projection lamp.

Material Required

- Flashlight;
- Discarded 35mm photographic slide;
- Wax paper;
- Rubber band;
- Hand lens;
- Tape.

Procedure

Cover the lens of a flashlight with a double thickness of wax paper. Be careful not to crumple or wrinkle this paper. (The wax paper lens will "soften" the bright centre of the beam).

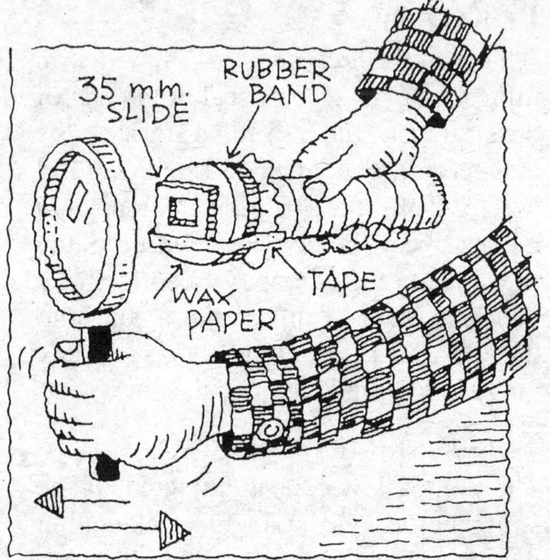

Pull the wax paper tight and secure this covering with a rubber band. Place a 35mm slide on top of the wax paper. Use tape to hold the 35mm slide in place.

Turn on the flashlight. Aim the beam at a light-coloured wall that is an arm's distance away. What do you see? Can you see colours? Is the image clear or fuzzy?

Place a hand lens several inches in front of the slide. Move the lens back and forth. Does the lens sharpen the projected image? Is the projected image larger or smaller than the actual drawn figure?

This is What Happens

You've built a film projector. Light that was emitted by the bulb streamed through the photographic slide. From there, the light continued outward, spreading as it travelled. By the time these rays struck the wall, the beam produced an out-of-focus image. To prevent the rays from spreading apart, a hand lens was placed into the flashlight beam. Its shape redirected the rays and focused them on the nearby wall, creating a sharp (and enlarged) image of the 35mm slide.

59. Shape Changing Bubbles

Soap bubbles represent a delicate balance of forces. Water tension creates the force that holds the thin film of the bubble together. Soap within the bubble solution offsets these forces and allows the bulb to stabilize. This results in a lightweight sphere whose shape is easily changed by static forces. Watch.

Material Required

- Bubble solution;
- Drinking straw;
- Mug or plastic container;
- Balloon.

Procedure

Fill a mug one-third full with bubble solution. Insert a straw into the solution. Blow into the solution with a slow and steady exhale. A mountain of bubbles should climb over the rim of the mug.

Charge a balloon by rubbing it through your hair. Move the balloon close to the bubbles. What happens? Describe how the bubble becomes distorted. Is the attraction strong enough to "rip" a bubble from the mug?

This is What Happens

Like the packaging "peanuts" and puffed cereal grains, soap bubbles react well to static charges. Their light weight and ease of charging make them an ideal subject for studying the effects of static attraction.

As the charged balloon approached the bubbles, the bubbles' near-by electrons reacted. These negatively charged particles migrated to the far side of the bubbles. This created a bubble top with a positive charge. The top was attracted to the negative balloon. This attraction caused the bubbles to stretch out and form an egg shape.

60. Making Electromagnet

In this easy experiment, we will make electromagnet or in other words we will try to find out how an electric current produces a magnetic field.

Material Required

- Wire, 18-gauge, insulated, 1 yd. (1 m);
- 6-volt battery;
- Long iron nail;
- Paper clips.

Procedure

Wrap the wire tightly around the nail, leaving about 6 in. (15 cm) of free wire on each end. (This wire-wrapped nail will be used in other experiments.) Have an adult strip the insulation off both ends of the wire. Secure one end of the wire to one pole of the battery. Touch the free end of the wire to the other battery pole while touching the nail to a pile of paper clips. Lift the nail while keeping the ends of the wire on the battery poles. When the nail starts to feel warm, disconnect the wire end you are holding against the battery pole.

This is What Happens

The paper clips stick to the iron nail.

Science Behind It

There is a magnetic field around all wires carrying an electric current. Straight wires have a weak magnetic field around them. The strength of the magnetic field around the wire was increased by coiling the wire into a smaller space, placing a magnetic material—the nail—inside the coil of wire, and increasing the electrical flow through the wire—attaching a battery. The iron nail became magnetized and attracted the paper clips.

61. Umbra and Penumbra

All shadows are created equal, or aren't they? Hold your hand several inches above this book. Raise and lower your hand while you observe your shadow. Does the appearance of the shadow change? If so, how? When is the shadow its sharpest? When is it darkest?

Material Required

- Scrap paper;
- Carbon paper;
- Pair of scissors;
- Flashlight;
- Wax paper;
- Rubber band.

Procedure

Place a sheet of scrap paper beneath the weird drawing on the next page. Then, insert a sheet of carbon paper (copying side down) between the page and the scrap. Trace out the odd outline.

Use a pair of scissors to carefully cut out the carbon copy's three irregular dark shapes.

As in the previous experiment, cover the lens of a flashlight with an unwrinkled, double thickness of wax paper. Pull the wax paper tight and secure this covering with a rubber band.

Hold the cutout about 2 inches away from the wall. Aim the beam of the flashlight at the cutout. What do you see? Are the edges of the image sharp or "soft"? How many shades can you distinguish in this shadow?

Slowly move the scrap paper away from the wall. What happens to the shadow? Keep the paper stationary and continue moving the flashlight back. What happens now?

This is What Happens

When the shadow mask is held close to the wall, a dark and sharp shadow is cast. People in science call this well-defined shadow an umbra. If you were the size of an ant and walked into this umbra, you wouldn't see the flashlight beam. It would be completely dark (except for the other light that was in the room).

As the shadow mask is moved away from the wall, lighter shades of shadow appear. These lighter shades form the shadow's penumbra.

The penumbra isn't in complete darkness. This time, our ant-sized visitor would see a part of the flashlight beam.

62. Making Switch

Take a look around you. Most likely, there's a nearby wall switch. Although you can't see it, the wall switch is connected to electrical wires that lead to a light fixture. When you flip the switch to the "on" position, the light goes on. Flip the switch "off" and the light magically goes off.

Material Required

- A small block of wood;
- Two thumbtacks;
- Metal paper clip;
- Two connecting wires with each end stripped 1 1/2 inches bare of insulation.

Procedure

Position two thumbtacks several inches apart on a small block of wood. Push them partially into the wood, leaving only a small space between the tacks' head and the wood surface.

Obtain two wires that have the insulation removed from both ends. Wrap a bare end from one of the wires around one of the thumbtacks. Push this tack into the wood to secure the wire.

Bend open a paper clip into an "S" shape. Slip one end of the paper clip's hook under the other thumbtack. Wrap this same thumbtack with the bare end of the second wire. Push this tack into the wood to secure both the wire and the paper clip.

Make sure that the paper clip extends over the other thumbtack. If not, bend it out some more. When you press down on the clip, it should make contact with the head of the first thumbtack. When you release, the clip should spring back up and "open" the circuit.

This is What Happens

The flow of electricity requires an uninterrupted path through which electric charges will move. The charges will stop flowing if there are any breaks in the path. It doesn't matter where the break is. As long as the path is not complete, no current will flow.

A switch is a device that can open and close a circuit. In the "on" position, the switch closes the circuit and completes the route for flowing current. In the "off" position, the switch places a gap in the path, which prevents any of the current from flowing.

63. Brain Trick

Brain trick. You are doing it right now. Although you are totally unaware of it, your brain is flipping everything it sees right-side up. Without this brain flip, your world would look as topsy-turvy as the images cast upon your retina.

But suppose an image was somehow projected right-side up? Would your brain know the difference and keep the scene as it appeared? Or would it perform a flip and turn the image upside down? Find out in this cool experiment.

Material Required

- Pin;
- Index card.

Procedure

Use the pin to poke a small pinhole in the centre of the index card. Stand in front of a brightly lit window. Hold up the card several inches from your face. Look through the hole at the bright outdoors.

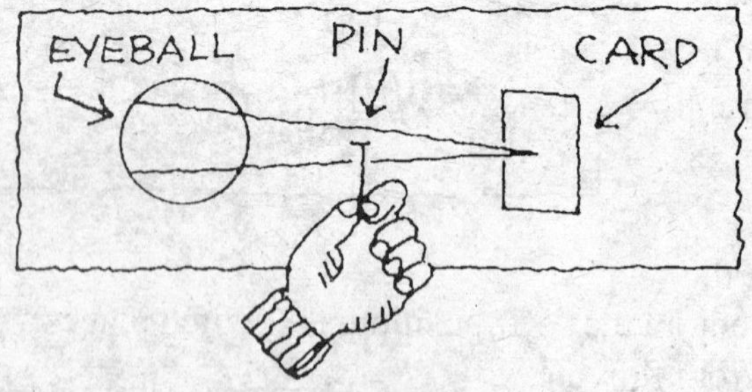

Hold the pin so that its flat head extends upwards. Position the pin head between the card's hole and your eye. First, hold the pin at a level just beneath the hole. Slowly raise the pin head. From which direction does it appear to cross the pinhole? Keep the pin head centred in the hole. Describe its appearance.

This is What Happens

Most of the time, the light rays that cast an image cross as they stream through the eye's pupil. This produces an upside-down view that your brain has learned to flip around.

In this experiment, you've created a unique setup. The rays that come through the pinhole won't cross over themselves. They only spread out. As they stream through the pupil, they remain uncrossed and cast a right-side up image on the retina. Your brain, however, has learned to flip over everything that it sees. So even though the image projected right-side up, your brain flips it upside down!

64. Drawing a Circuit

Although this next experiment involves drawing, don't worry. You don't have to be an artist. All you have to do is copy simple symbols and arrange them to represent electric circuits. So pick up your pencil and give it a try!

Material Required

- Pencil;
- A sheet of paper.

Procedure

Examine the symbols below. They represent different parts of a circuit. Can you figure out what the complete circuit looks like? Look at the next page to find out.

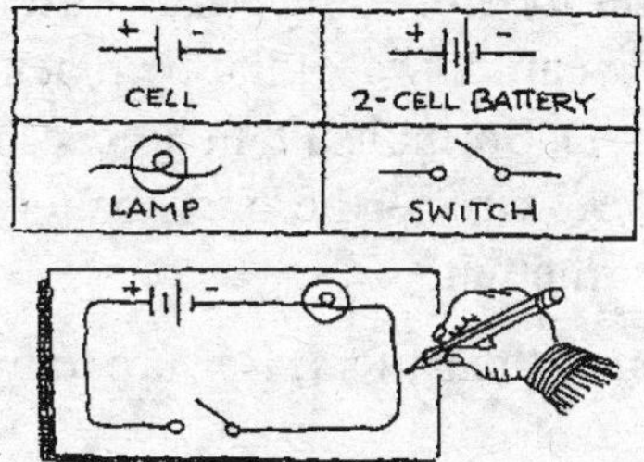

Now it's your turn. Use these symbols to draw diagrams that represent the following circuits. Don't forget about the polarity of each cell and how to illustrate it correctly.

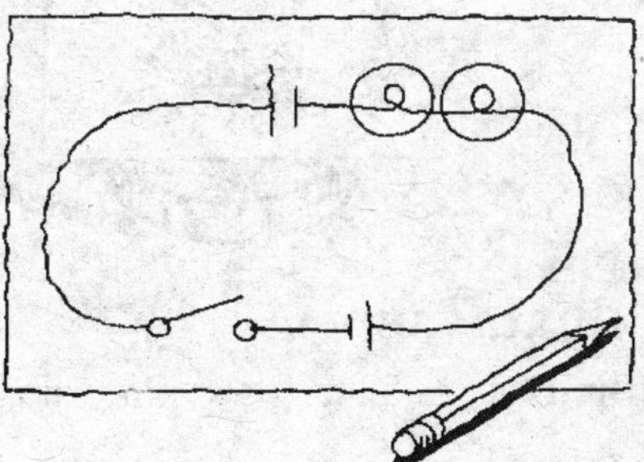

This is What Happens

Symbols are awesome. By the way, the answer is:

65. Starburst Pattern

The intention of this experiment is to demonstrate how electricity and magnetism are related.

Material Required

- Wire-wrapped nail from Electromagnet experiment;
- 6-volt battery;
- Masking tape;
- Cardboard, 6 in. (15 cm) square;
- Iron filings;
- Scissors.

Procedure

Ask an adult to punch a hole through the centre of the cardboard, using a nail. Insert the wire-wrapped nail through the hole in the cardboard. Make the cardboard sit flat by placing it on the roll of tape. Attach one end of the wire to either battery pole. Sprinkle a thin layer of iron filings on the cardboard around the coiled wire. Attach the free wire to the open battery terminal. Observe the pattern made by the iron filings. Disconnect the wire.

Caution: The nail and wires will get hot if left connected to the battery. Be sure to interrupt the circuit by disconnecting one wire from one pole.

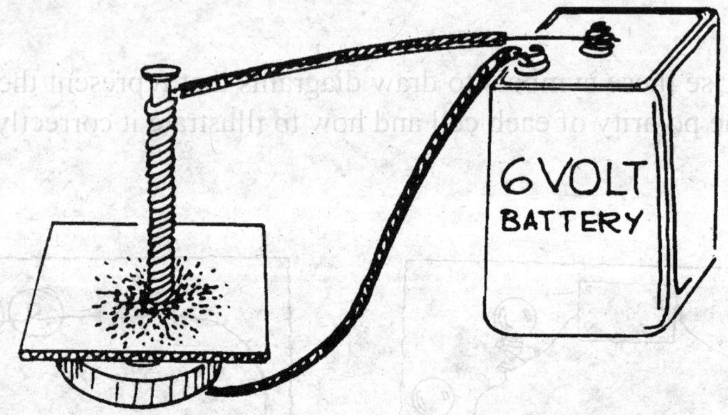

This is What Happens

The iron filings form a starburst pattern around the coil of wire.

Science Behind It

There is a magnetic field around all wires carrying an electric current. The magnetic effect around the wire can be increased by coiling more wire into a smaller space, increasing the electrical flow through the wire, and placing the iron nail in the coil of wire. The iron filings are pulled towards the magnetized nail and form a starburst pattern around the coil of wire.

66. Benham's Disk

Suppose you mixed blue and yellow pigments. You'd expect to see green.

Suppose you mixed red and white pigments. You'd expect to see pink. But suppose you looked at a spinning disk that had a black-and-white pattern. What colour or shade would you expect this blur to be? Make a guess. Then be prepared for a surprise.

Material Required

- Heavy stock paper;
- Pair of scissors;
- Glue stick;
- Paper clip;
- Pushpin.

Procedure

Make a photocopy of the disk on the opposite page. If possible, print out this copy on heavy stock paper. If stiff paper stock is not available, you'll need to support the photocopy with a heavy stock backing. To make this backing, cut out the photocopied disk. Trace this disk onto a sheet of stiff paper. Cut out the stiff paper disk and use a glue stick to paste it to the backside of the photocopy.

Unbend one of the "arms" of a paper clip so that it looks like this.

Use the pushpin to punch a hole through the centre of the disk. Place the bent-up arm of the clip through this hole. The disk should rest flat and spin freely on the clip bend. Hold the disk beneath a bright light. Spin it and observe what happens to the distinct black and white pattern. Does this odd effect change as you vary the speed of the spin?

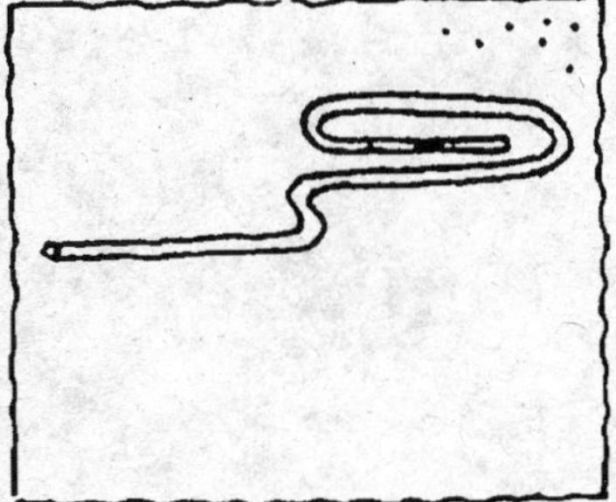

This is What Happens

Colours? Where did those colours come from? No one seems to know for sure, although scientists have made some pretty good guesses.

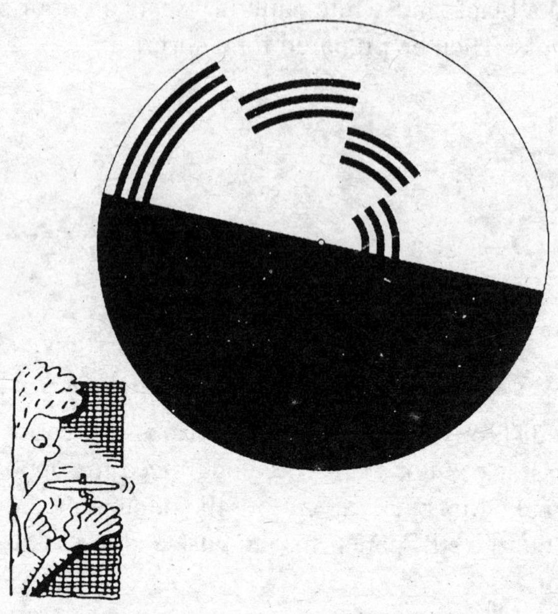

This black-and-white pattern is called a Benham's disk. It was first created over 100 years ago! When it spins at the right speed, it produces a changing pattern of light that falls upon your eye's retina. Many scientists think that the pattern creates a nerve message that assembles the message for colour. So when this message (which should "code" for a black-and-white spin) arrives at the brain, its content is misinterpreted. Instead, your brain thinks that this message describes a colour. And so, colours where there are no colours!

67. Pencil Dimmers

Dimmers are great devices. Some of these circuit components are attached to ceiling lights or floor lamps. As you rotate the dimmer's knob, the light gets brighter or dimmer.

Material Required

- Pencil (or just a piece of pencil lead AKA graphite);
- Two "D" cells in battery holders;
- One flashlight bulb in lamp holder;
- Connecting wires with each end stripped 1 1/2 inches bare of insulation.

Procedure

Have an adult use a wood-carving tool to shave down one side of a pencil so that a long length of graphite is exposed. It should look like this:

If you can't find an adult to do the carving, do not do it yourself! Instead, you can use a piece of graphite that's meant for use in refillable pencils.

Assemble the circuit below.

Touch the bare ends of the two wires together and the lamp should light. If not, check your connections. Now place the ends of both tester wires against the graphite. What happens? Slowly move these wires farther apart along the length of graphite. What happens to the brightness of the lamp? Can you explain your observations? Predict what will happen if the wires are moved together. Find out if your prediction is correct.

This is What Happens

Unlike copper, graphite isn't a good conductor of electricity. When electricity flows through graphite, it crashes "head-on" into resistance. This resistance cuts down on the flow of electricity.

When your circuit included only a small section of graphite, the lamp remained bright. As the wires were drawn apart, the current was forced to travel a greater distance through the graphite path. This extra distance produced increased resistance, which cut down the electricity and made the bulb dimmer.

68. Magnetic Attraction

Are all types of material attracted to a magnet? In this experiment, we will find out which material is attracted and which is not.

Material Required

- Magnet coins: penny, nickel, quarter, dime;
- Iron nail;
- Paper clip;
- Aluminum foil, small coin-sized piece;
- Pencil paper.

Procedure

Touch the magnet to every object. The pencil has several parts: eraser, metal band, wood, and pencil point. Be sure to test all parts. Make a list of the items that are attracted to the magnet.

This is What Happens

The nail and paper clip are the only items attracted to the magnet.

Science Behind It

Most materials are not attracted to a magnet. Those that are can become magnetized themselves. Magnetic materials contain magnetic domains, clusters of atoms that behave like tiny magnets. When the domains are randomly arranged, the material is not magnetic, but if the domains are all lined up, the material has magnetic properties. The steel paper clip and iron nail contain magnetic domains that were in random order before being touched by the magnet. The magnet's magnetic force pulled on the magnetic domains in the nail and paper clip, causing them to point towards the magnet. The lining up of the domains magnetized the materials. The magnet and the now-magnetized paper clip and nail are attracted to each other.

69. Strobe Machine

There is a limit to the amount of information that your brain can process. Think of a cartoon flipbook. When you observe a rapidly flipped sequence of drawings, it becomes impossible to distinguish individual frames. Instead, your brain gets fooled into "seeing" the illusion of continuous action.

In this experiment, you'll construct a type of strobe machine. Unlike the dance floor lights, this device doesn't depend upon flashes to create its visual trickery. Instead, the motion capture is produced by quick glimpses through rapidly spinning notches.

Material Required

- Pair of scissors;
- Pencil with eraser;
- Pushpin;
- Television screen;
- Kitchen sink (and faucet);
- Mirror.

Procedure

PART 1: Hold That Pose

Make a photocopy of the disk on the opposite page. Use a pair of scissors to cut out the disk notches. Insert a pushpin through the centre of the disk and anchor it to the side of a pencil's eraser. The disk should spin freely. Make sure that you assemble this spinner so that the images face away from the pencil.

First, we'll use this strobe viewer to freeze motion. Turn on a television or computer screen. Stand across the room from the screen. Shut one eye and look through any of the notches at the screen. Spin the disk. Keep looking.

As the notches rotate into view, you'll catch a quick glimpse of the screen. Does the screen's appearance change? How does increasing the spin speed affect the screen's appearance?

Adjust a cold water faucet so that a slow but steady trickle of droplets fall into the sink. Observe this motion through the notches of the spinning disk. Can you freeze the droplets in midair?

If you are having trouble seeing the droplets, try this. Fill a large paper cup with water. Add several drops of dark food colouring. Over a sink, have a friend use a pushpin to punch a hole in the bottom of the cup. Use your spinning disk to observe this stream of coloured droplets.

PART 2: Movie Magic

Stand in front of a mirror. Look through one of the notches. Observe the reflected pattern of horses. Spin the disk. Keep looking through the notches. What do you see?

This is What Happens

Motion that you observe on a TV screen (or movie theatre or flipbook) is an illusion. It is created by a series of rapidly changing frames. Onscreen, these electronic images get "redrawn" 30 times per second. At this rate, your brain can't distinguish individual frames. In its confusion, your brain produces the illusion of smooth motion.

This strobe viewer can capture a quick glimpse of partially refreshed screens. The black bands that you observe are part of a series of incomplete screens, all captured at the same stage of refreshing.

Like the "refreshing screen," the falling droplets can also appear frozen. To produce the freeze, each quick glimpse must catch the next droplet in the same position as the previous one. This produces the illusion of a single droplet that remains in midair. To make the droplet appear to go up, each glimpse must catch the next droplet at a slightly higher position than the previous one.

Your strobe device can also be used to produce the illusion of animation. As the disk spins in front of a mirror, it offers a quick glimpse the reflected horse images. At a fast enough speed, your brain can't distinguish the individual drawings. Instead, it blends the sequence of rapidly changing images into the illusion of smooth, animated motion.

70. Short Circuits

Shocking Question: What would happen if you were to connect the positive and negative terminals of a cell together without placing a bulb, motor, or some sort of resistance in the circuit?

Answer: If the circuit were left on for more than a few minutes, you'd ruin the cell (and if the current source was large enough, you'd burn yourself or start a fire).

Material Required

- Two "D" cells in battery holders;
- One flashlight bulb in lamp holder;
- Two switches;
- Connecting wires with each end stripped 1 ½ inches bare of insulation.

Procedure

Assemble the circuit below.

Close only switch A. What happens to the lamp? Is there a closed circuit? If so, trace the complete circuit through which electricity flows.

Close only switch B. What happens to the lamp? Is there a closed circuit? If so, trace the complete circuit through which electricity flows.

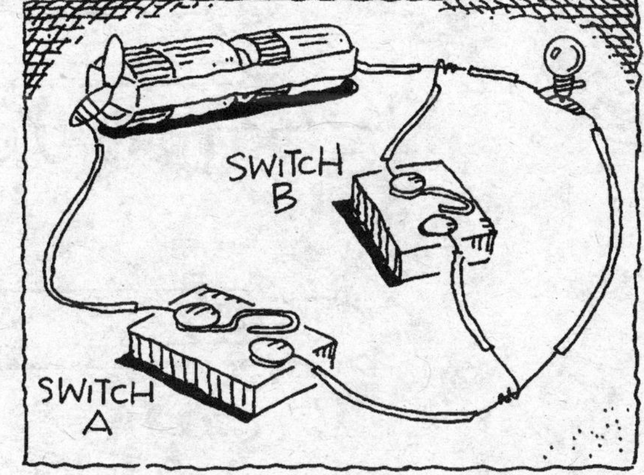

Bend the clip in switch A so that this switch remains on. While the lamp is glowing, close switch B. What happens? What path does the current flow when both switches are closed? Why? Release switch B. What happens now? Why?

This is What Happens

Switch A controls current flow in the outer loop. When switch A is closed, current flows along this outer path to light the lamp.

Switch B is part of a separate path that offers no resistance. When switch B was closed, its path became complete. Since the path has no resistors, it acted as a short circuit. The current flowed across switch B and bypassed the lamp. Without a current flow, lamp went out.

71. Metal Effects

The purpose of this easy experiment is to determine how metals affect a magnetic field.

Material Required

- Aluminum foil;
- Steel spatula;
- Bar magnet;
- 4 small paper clips.

Procedure

Lay the paper clips on a table and cover them with a sheet of aluminum foil. Set the magnet on the foil over the clips. Raise the magnet and observe any movement of the clips. Position the clips so that they lay under the spatula. Set the magnet on top of the spatula. Lift the spatula with the magnet and observe any movement of the clips.

This is What Happens

The magnet attracts the paper clips through the aluminum foil. The magnet does not attract the paper clips through the steel spatula.

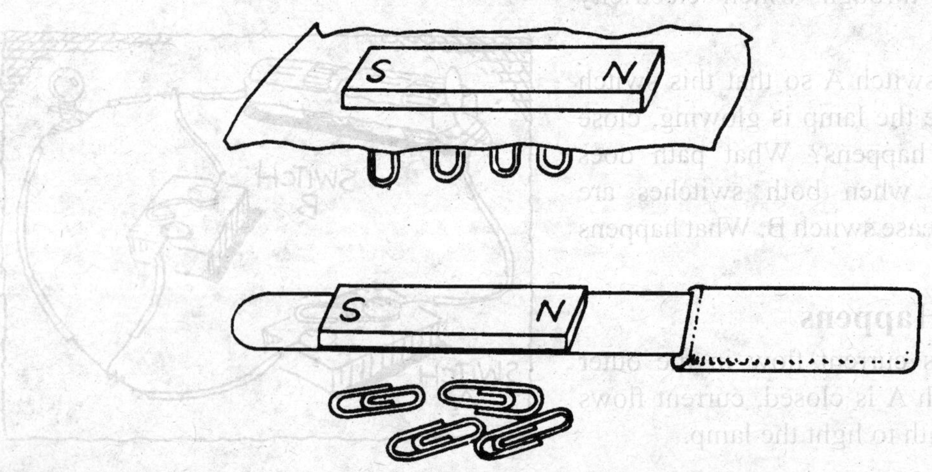

Science Behind It

The magnetic force field passes through the aluminum, but the steel blade restricts the movement of the force field. The steel blade is attracted to the magnet, but the metal provides another path for the magnetic field. This new path is through and around the steel blade. The steel keeps the lines of force closer to the magnetic field, acting as a barrier to other magnetic materials.

72. Dancing Puffs

If you have seen some war movies, you have felt an on-screen explosion? If so, you know that sound can be a moving experience. Do you remember which sounds shook you the most? Was it the high pitched whirls or the booming low pitches? While you are remembering, here's a smaller-scale rumbling (but one that's easy to observe).

Material Required

- Sound system;
- Foil pie pan;
- Puffed cereal grains.

Procedure

Gently lay a speaker cabinet on its back so that its speaker faces upwards. Balance a foil pie plate on top of the speaker grill.

Pour about fifty grains of puffed cereal onto the pie plate. Make sure that the sound system is "off," with the volume turned down. Switch on the system. Slowly crank up the volume and watch what happens to the cereal!

This is What Happens

Sound is a form of energy. It consists of vibrations that are passed from particle to particle. In air, the vibrations of sound are transferred by particles of gas that are too small to see.

The sound produced by the speaker caused air particles to vibrate. The energy of the vibration spread out and struck the bottom of the pan. The large (and light) bottom of the pan was struck by a good deal of sound energy. This energy caused the pan to vibrate. The energy of vibration was then transferred to the cereal grains. As they became energized, the puffs jumped to the beat.

73. Series Circuit

Have you ever tested old batteries in a 2-cell flashlight? If so, you know that both of the cells need to be working. If one cell has leaked, the flashlight won't work. No matter how strong the other cell is, if one cell is bad the light won't go on. That's because the electricity needed to light the bulb must flow through both cells. If one cell is dead, the path is blocked. There's no way around it.

Material Required

- Two "D" cells in battery holders;
- Two flashlight bulbs in lamp holders;
- Switch;
- Connecting wires with each end stripped 1½ inches bare of insulation.

Procedure

Assemble the setup shown below.

Close the switch. What happens? Do both bulbs light? Is one bulb brighter than the other? In this circuit, how many different paths can the electrons flow through?

Keep the switch closed. Unscrew one of the bulbs. What happens?

Can you explain your observations? Draw a diagram of this circuit using the symbols presented earlier. Use this diagram to describe the flow of electrons.

This is What Happens

The circuit you built is called a series circuit. It offers only one path for the moving charges. All the electrons must travel through the same circuit components.

If one of the wires is broken (or one of the components removed), the circuit is opened. All current stops! That's why the other bulb went out, even though it remained attached to the circuit.

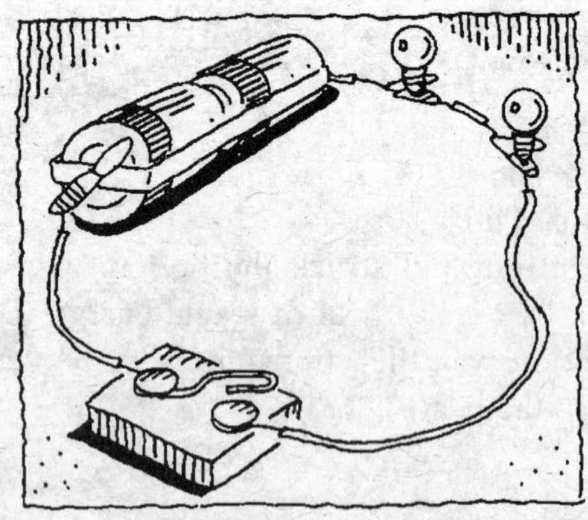

74. Floating Bottle

How salt affects buoyancy? In this experiment, we will try to understand this phenomenon.

Material Required

- Large-mouthed jar, 1 gal. (4 litre);
- Small bottle with a lid table salt, 2 cups;
- Measuring cup, 1 cup (250 ml);
- Large stirring spoon.

Procedure

Fill the jar three-quarters full of water. Carefully place the closed small bottle in the water. It should float on the surface. If it does not, get a smaller bottle. Remove the small bottle and add a small amount of water. Close the lid and replace the bottle in the gallon jar of water. The bottle should slowly sink to the bottom. Remove or add water to the small bottle until it sinks at a slow rate when placed in the water. Remove the bottle and add ½ cup of salt to the large jar of water. Stir. Replace the small bottle in the jar and observe its position in the water. Continue to add ½ cup of salt at a time until 2 cups have been added. Observe the position of the small bottle in the salty water after each measure of salt has been added.

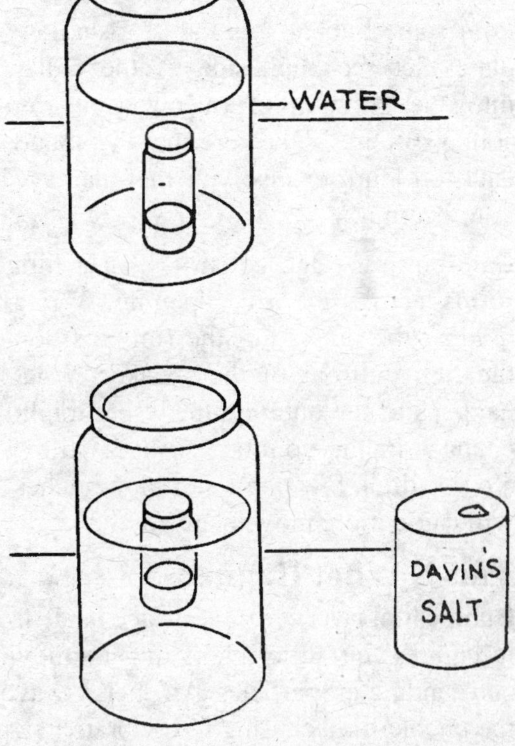

This is What Happens

The bottle rises in the water as more salt is added.

Science Behind It

Adding water to the small bottle makes it heavy enough to sink. The sinking bottle pushes water out of its way. The weight of the water pushed aside equals the amount of upward force on the bottle. This force is called the buoyancy force. A small amount of salt increases the buoyancy force, but it is not enough to overcome the downward force resulting from the weight of the bottle and water inside. The larger amount of salt increases the buoyancy force, allowing the bottle to float on the surface of the heavy salty water.

75. Filmy Sound

Have you ever removed the front cloth or grill covering of a speaker cabinet? If so, you exposed the sound-pushing surface of the speaker. Turn up the stereo and the speakers vibrate back and forth. The louder the sound, the greater the movement. In this experiment, you'll construct a device that will also move with the music. Unlike the plastic or cardboard speaker cone, your device will be made from something slightly more fragile—soap bubble film!

Material Required

- Paper drinking cup;
- Bubble solution;
- Pair of scissors;
- Speaker;
- A shallow plate.

Procedure

Use your scissors to cut off a "collar" from the cup's rim. The collar should be about 1 inch wide and form a continuous round band.

Pour some bubble solution in a shallow plate. Set the "lip" side of the collar into the solution. Turn on a nearby sound system. Make sure that the sound isn't too loud, or else you may damage your hearing.

Carefully raise the collar so that a film forms across the rim opening. While holding the outside of the rim, position the film in front of the speaker. What happens to the bubble film? How might changes in the volume affect the film? Do you think low pitches or high pitches will cause more movement? Why?

This is What Happens

Bubble film is a very thin surface made from water and soap particles. Since it is lean and light, it requires very little energy to get it moving. As the speaker cone moved in and out, it vibrated surrounding air particles. An energy wave was formed. As this wave moved outward, it struck the bubble film, causing it to vibrate in a pattern that copied the sound's intensity.

76. Parallel Circuit

Have you ever been to a party that was lit up by strings of overhead lights? Sets of these lights hang as holiday or patio decorations. As you may have observed, some of these light sets are hooked up in a special way. When one bulb goes out, the others remain lit! This way of wiring makes it much easier to identify the burned-out bulb. Otherwise, you'd be faced with testing the entire set.

Material Required

- Two "D" batteries in battery holders;
- Two flashlight bulbs in lamp holders;
- Connecting wires with each end stripped 1½ inches bare of insulation;
- Switch.

Procedure

Assemble the setup shown below.

Close the switch. What happens? Do both bulbs light? Is one bulb brighter than the other? In this circuit, how many different paths can the electrons flow through?

Keep the switch closed. Unscrew one of the bulbs. What happens? Can you explain your observations?

Draw a diagram of this circuit using the symbols presented earlier. Use this diagram to describe the flow of electrons.

This is What Happens

The circuit you built is called a parallel circuit. It offers more than one path for the moving charges. This circuit had two possible paths. The current of flowing electrons split up. Half the current travelled through the nearer "arm" of the circuit. The other half travelled through the distant "arm."

This time, when a bulb was removed, the remaining bulb stayed lit. That's because the path to the untouched bulb remained intact. The electrons continued to flow along this smaller but complete circuit loop.

77. Bubbles Up

Why bubbles rise in liquids? This is what we are going to find out.

Material Required

- Large-mouthed jar, 1 gal. (4 litre);
- Clear plastic tubing, 20 in. (50 cm);
- Small balloon.

Procedure

Fill the jar with water. Place one end of the clear plastic tubing in the water at the bottom of the jar. Inflate the balloon and twist the neck to prevent the air from escaping. Slip the mouth of the balloon over the end of the tube. Hold securely with your fingers. Untwist the balloon and allow the air to escape slowly through the tube. Watch the end of the tube in the water and notice the movement of the air as it exits in the tube.

This is What Happens

Bubbles are formed at the end of the tube. The bubbles rise to the top of the water's surface and escape into the air.

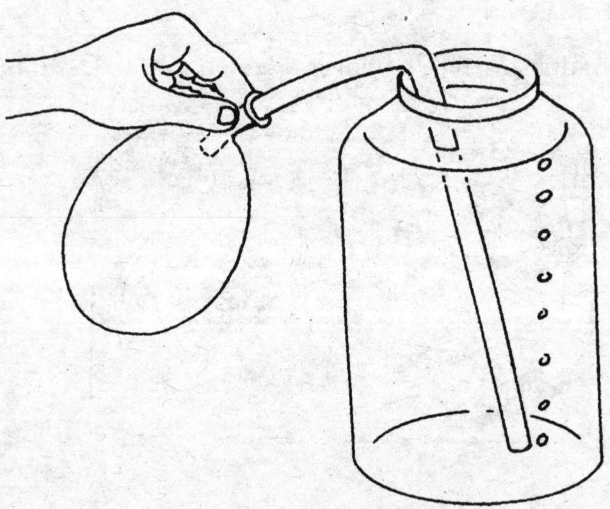

Science Behind It

The gas bubbles push water out of their way as they emerge from the end of the aquarium tube. The weight of the water pushed aside equals the amount of upward force on the bubbles. This force is called the buoyancy force. The gas bubbles are so light that they quickly push to the top of the water where they break through the water's surface and mix with the air surrounding the jug.

78. Spotting Sound

Place your palm about 1 foot in front of your mouth. Count from one to ten. As you count, slowly bring your palm closer to your mouth. Most likely, by the time you reach "five" you'll feel the sound waves as they strike your hand. Now keep you palm about 6 inches from your mouth. Again, count from one to ten. Can you "feel" the difference in the sounds of each number?

Material Required

- Cardboard tube (from a finished roll of toilet paper);
- Wax paper;
- Rubber band;
- Sequin or mirrored plastic chip;
- Glue;
- Pair of scissors;
- Flashlight.

Procedure

Cut out a 4-inch-square wax paper. Stretch the paper over one end the cardboard tube. Pull it tight to form a flat surface. Slip a rubber band over the wax paper to keep it in place.

Use glue to secure a sequin to the centre of the wax paper. Hold the open end of the tube to your mouth. Use your other hand to aim flashlight beam at the sequin. Stand near a wall so that you can locate the reflected spot of light. Talk loudly into the tube. Observe how you or voice affects the spot.

This is What Happens

As you spoke, you produced sound waves. These waves (which consist of pushed-together particles of air) moved outwards from your mouth and travelled into the tube. At the far end of the tube, the waves struck the wax paper. This collision caused the stretched paper to vibrate. As the paper moved, the spot of light reflected by the sequin traced out the vibration pattern.

79. Basic Buzzer

It's time for a break. Instead of resting, however, you get to assemble a simple circuit. This working circuit will be used in several of the following experiments. Right now, you need to find the components, make sure they match up, and assemble the circuit correctly.

Material Required

- Buzzer or bicycle horn (1.5-volt or 3-volt device);
- One or two "D" cell batteries;
- Battery holders;
- Connecting wires with each end stripped 1½ inches bare of insulation;
- Wire stripper.

Procedure

Find a buzzer or bicycle horn. Buzzers can sometimes be removed from old board games that have "buzzing" sounds. If you do remove the buzzer from a game, write down the type and number of cells needed to energize the buzzer. Remember, you'll need to match your battery supply to the electrical needs of this device.

If you don't find a board game buzzer, a bicycle horn might work. Some horns can be removed from their streamlined case. If so, you'll need to identify the terminals to the horn. If the horn can't be separated from the case, remove the cell. Attach wires to the terminals that are exposed inside the battery holder. These wires can then be attached to your circuit as direct paths to the buzzer. Again, you'll need to match up the horn with the correct power supply.

And if all else fails, there's always the local electronics store. They should have both 1.5-volt and 3-volt buzzers.

Your simple circuit should look like what's shown on the following page. In the following experiments, you'll need to add this circuit to different types of switches. Before you do, make sure that the buzzer works.

This is What Happens

When the circuit is complete, electricity flows. This current of electrical charges "energizes" the horn, which causes its sound-producing parts to vibrate. We detect the vibrations as a buzzzz.

80. Floating Boat

You may have seen large ships made of metal or other materials floating in water instead of sinking. In this experiment, we are going to find out how a heavy ship floats.

Material Required

- 20 paper clips;
- Aluminium foil;
- Ruler;
- Bucket of water.

Procedure

Cut two 12 in. (30 cm) squares from the aluminium foil. Wrap one of the metal squares around 10 paper clips and squeeze the foil into a tight ball. Fold the four edges of the second aluminium square up to make a small square pan. Place 10 paper clips in the metal pan. Set the metal pan on the water's surface in the bucket. Place the metal ball on the water's surface.

This is What Happens

The metal pan floats and the ball sinks.

Science Behind It

The ball and pan both have the same weight, but the ball takes up a smaller space than does the pan. The amount of water pushed aside by an object equals the force of water pushing upward on the object. The ball pushes less water out of its way than does the larger pan, and thus there is not enough upward force to cause it to float. Large ships are very heavy, but they have hollow compartments filled with air, which increases their buoyancy.

81. Sound and Matter

In order to produce sound, you need something to vibrate. Matter is that something. It's the "stuff" that the universe is made of. But will all matter transmit sound? Are there only certain particles that can pass these sound waves along? Let', find out.

Material Required

- Three plastic sealable bags;
- Sand (sugar or flour can be substituted) water;
- Coin.

Procedure

Fill the first bag halfway with sand. Push out the extra air and seal the bag shut. Fill a second bag halfway with water. Again, push out the extra air and seal the bag shut. Make sure that it doesn't leak. Fill third bag halfway with air. Seal it and place all three bags on a table top.

Place your ear on the bag of sand. Gently tap a coin on the tabletop. What do you hear? Repeat the coin taping while listening through the bag of water. Does it sound the same? How has the taping sound changed? Repeat the tapping one more time while listening through the bag filled with air. Compare the three sounds.

This is What Happens

Sounds are transferred by particles that slam into neighbouring particles. Solids contain particles with the closest neighbours. Next come liquids. Gases have the most spread-out particles.

The closer the particles are to each other, the better the sound transmission. That's why the sound that passed through the solid and liquid may have seemed much louder than the sound that passed through the air filled bag.

82. Alarm Device

Alarm systems are great, especially the high-tech ones that detect an intruder by the heat he or she gives off. In this experiment, you'll get to construct an alarm device and use it to assemble a circuit.

Material Required

- Two connecting wires with each end stripped 1½ inches bare of insulation;
- Simple buzzer circuit;
- 10 inch x 3 inch cardboard;
- Pair of scissors;
- 2-inch strip of aluminum foil;
- Tape.

Procedure

Cut out a strip of cardboard (removed from a packing box) that is about 10 inches long x 3 inches wide. Bend the strip in half to form two 5-inch-long sections. Wrap the middle of each section with a 2-inch strip of aluminum foil. Use tape to secure the strip. Attach two 1-foot long wires to the outer side of each strip. These wires are attached to a series circuit that contains two 1.5-volt cells and a buzzer. The folded cardboard should be placed beneath a carpet or mat. Make sure that the weight of the carpet doesn't close the halves together. Once the alarm detector is "set," just sit back and wait.

This is What Happens

The alarm detector is a pressure switch. The weight of the carpet is not great enough to fully press the cardboard halves together therefore, the switch remains open. When someone stands on the carpet, however, their added weight closes the switch. Once closed, the current can flow from the battery to the buzzer and sound the alarm!

83. Rising Clay Balls

The purpose of this easy experiment is to determine how the buoyancy of a substance can be changed.

Material Required

- Drinking glass;
- Club soda;
- Modeling clay.

Procedure

Fill the drinking glass three-quarters full with soda. Immediately add 5 tiny balls of clay one at a time. The clay pieces must be about the size of a rice grain. Wait and watch.

This is What Happens

Bubbles collect on the clay. The clay pieces rise to the surface, spin over, and fall to the bottom of the glass, where more bubbles start to stick to them again.

Science Behind It

The soda contains carbon dioxide, which forms bubbles that stick to the clay. The clay balls initially sink because their weight is greater than the upward buoyant force. The gas bubbles act like tiny balloons that make the balls light enough to float to the surface. The carbon dioxide bubbles are knocked off at the surface, and the balls again sink to the bottom until more bubbles stick to them.

84. Surface and Sound

Let's be critical. Fill a plastic bag full of water. Seal the bag and listen through it by placing your ear against its surface. Which type of matter did the sound have to travel through in order to reach your ear?

a) solid; b) liquid; c) gas

If you answered "liquid," you're only partially correct. The sound was also carried by the solid plastic covering.

Material Required

- Wristwatch (old wind-up type);
- Large plastic sealable bag.

Procedure

Wind up the wristwatch. Make sure that it works. Place the watch on a flat surface. Position your ear about 6 inches above the watch. Can you hear it tick?

Fill a plastic bag with air and seal it shut. Place the bag over the watch. Place your ear on the surface of the bag. Can you hear the ticking now? Why?

This is What Happens

Without the bag, the ticking sound was transferred only by air. Since air particles are very spread out, this type of transfer is not effective.

When the bag full of air was used, things changed. The ticks were absorbed by the plastic skin of the bag. The sound energy was then transferred by the plastic particles along the surface of the bag. Since the particles of this solid were concentrated, less sound energy was lost. This produced a volume that was loud enough to detect.

But don't forget, the total sound also depended upon what was in the bag. In this case, it was air.

85. Making Door Alarm

In movies, you have seen when an intruder opens a door, the alarm goes on. In this experiment, you can make one for yourself.

Material Required

- Connecting wires with each end stripped 1½ inches bare of insulation;
- Simple buzzer circuit;
- 5 inch x 3 inch cardboard;
- Pair of scissors;
- Metal paper clip;
- Tape;
- Two metal thumbtacks;
- String;
- Small square piece of paper.

Procedure

Cut out a strip of cardboard (removed from a packing box) that is about 5 inches long x 3 inches wide. Push two metal thumbtacks into the cardboard. The tacks should be about 2 inches apart. Wrap a length of conducting wire around the shaft of each tack.

Unbend a paper clip into an "S" shape. Slip one end of the clip around the base of one tack. Adjust the bend in the clip to ensure that there is enough pressure to keep the far end of the clip pressed down on the other tack.

Slip a small square of paper between the clip and the tack. The paper will be held in place by the pressure of the bent clip.

Tape a string to the paper. Tie the other end of the string to a doorknob. Attach the conducting wires to a series circuit that contains two 1.5-volt cells and a 3-volt buzzer.

Tape the cardboard setup down so that it remains secure. Now invite a friend to open the door. BUZZZZZZZZZZZZZ.

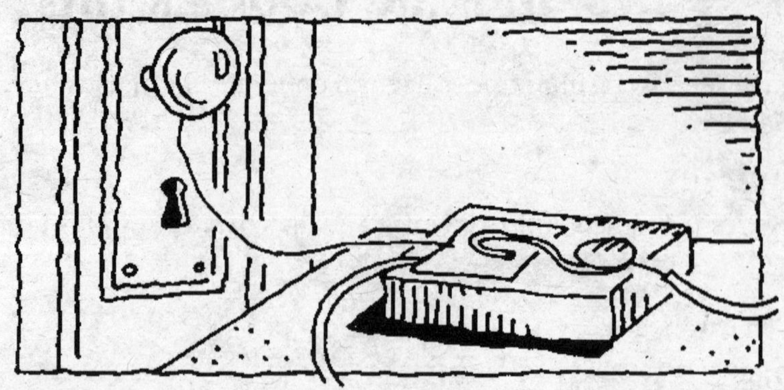

This is What Happens

The paper that was inserted between the clip and the tack acted as an insulator. This insulator blocked the flow of electric current. When the paper was yanked away, the clip and tack made contact. When the circuit became complete, charges flowed and the alarm sounded.

86. Rolling Funnels

Purpose of this experiment is to determine the effect that an object's centre of gravity has on motion.

Material Required

- 2 yardsticks (metre sticks);
- 3 books, each at least 1 in. (2.5 cm) thick;
- Masking tape;
- 2 funnels of equal size.

Procedure

Put two books 30 in. (90 cm) apart on the floor. Place the remaining book on top of one of the other books. Position the yardsticks on top of the books to form a V shape with the open part of the letter on the double book stack. Tape the bowls of the funnels together. Place the joined funnels at the bottom of the track formed by the yardsticks.

This is What Happens

The funnels roll up the hill.

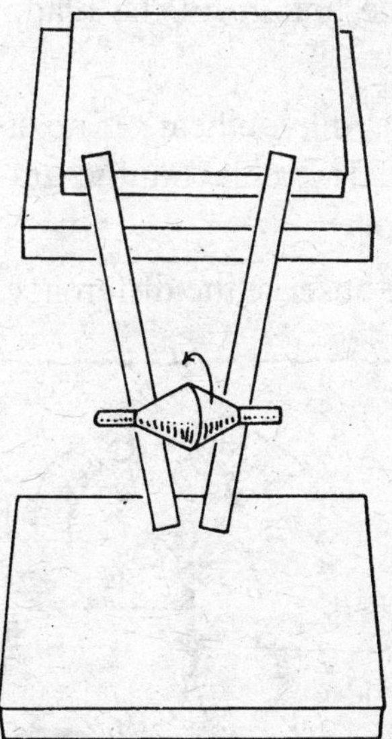

Science Behind It

The funnels are not defying the laws of gravity. Actually, as the joined funnels move, their centre of gravity (the point where the weight appears to be located moves downward. Notice that the centre of the joined funnels gets closer to the floor as it moves along the raised yardsticks.

87. Sounds from Grass

If you have ever looked closely at a saxophone, clarinet, mouth-piece or shahnai? If so, you probably observed that the mouthpiece contains a thin piece of wood. This part of the instrument is called a reed. When you blow across the reed, you set it vibrating. It's this back-and-forth movement that produces the instrument's sound. As these vibrations travel through the instrument's body, the sound is "shaped" into a pleasing note.

Material Required

- Blade of grass;
- Paper;
- Pair of scissors;
- Clear plastic wrapping.

Procedure

Hold out your hands. Place your thumbs side-by-side together. You should observe a slender gap between them. This opening begins at the middle joint and stretches downwards to the base of the thumbs.

Now position a blade of grass in this passageway. The top of the blade should be held in place between the two middle joints. The bottom of the blade should be secure between the two thumb bases.

Blow through the passageway. What do you hear? If no sound is produced, make sure that the blade lies in the middle of the gap. Try replacing the grass blade with a blade-sized scrap of paper (you may have to wet the paper).

Practice with various materials and observe the difference in the quality of the sounds.

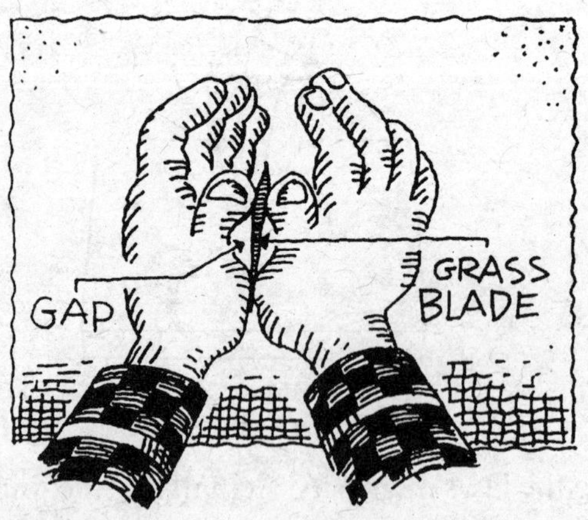

This is What Happens

As air rushed through the slender opening, it caused the blade to vibrate. This vibration shaped the passing air into sound waves.

88. Testing Your Steadiness

Have you ever played the electrified board game in which you must operate on a patient? If your hand isn't steady, the forceps strike the conducting metal. Instantly a loud sound confirms that you "grounded out" and completed your turn. In this experiment, you'll have the chance to build another type of game that also tests your steadiness.

Material Required

- Metal clothes hanger;
- Simple buzzer circuit;
- Two lumps of clay;
- Connecting wires with each end stripped 1½ inches bare of insulations;
- Steel wool.

Procedure

Use steel wool to remove any sort of coating or lacquer that covers the clothes hanger.

Untwist the coat hanger and rebend it into a large "U". Make several bumps in the middle section of the "U". Stand up the ends of this clothes hanger with lumps of clay.

Assemble the battery of two "D" cells. Attach a connecting wire from one base of the clothes hanger to one terminal of the buzzer. Attach another connecting wire from the other terminal of the buzzer to one of the ends of the battery. Now attach the 2-foot-long connecting wire to the other end of the battery. Bend the free end of the 2-foot wire into a small loop that fits around the coat hanger.

The object of this game is to move the loop along the entire path and not make contact with the hanger.

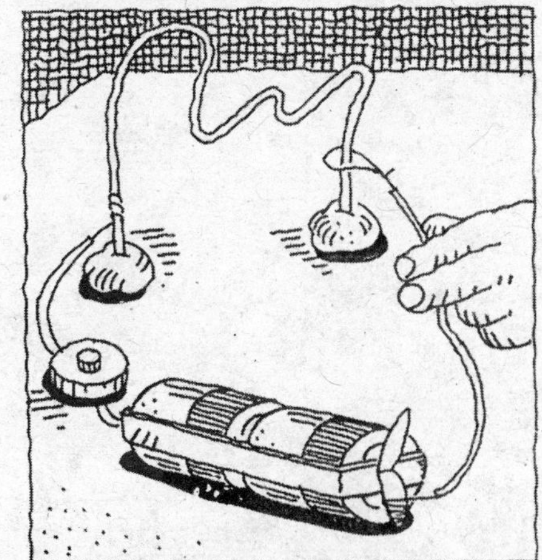

This is What Happens

In order to "win", you have to maintain an open circuit. If your hand is not steady, the loop will make contact with the conductive hanger. Once contact is made, the circuit becomes complete and the buzzer sounds to announce your defeat!

89. Bigger the Better

Parachute is an umbrella-like device for slowing the descent of a body falling through the atmosphere. Separate panels sewn together form a canopy attached by suspension lines to a harness worn by the user. Originally designed to provide a safe escape from a disabled aircraft, parachutes are also used for dropping supplies and for slowing returning space capsules.

In this experiment, we will find out if size affects the speed of a falling parachute.

Material Required

- Small plastic garbage bag;
- String 2 small;
- Washers;
- Scissors;
- Ruler.

Procedure

Cut eight separate strings about 20 in. (50 cm) long. Cut a 12 in. (30 cm) square from the plastic. Tie a string to each comer of the plastic square. Tie the four free ends of the strings together in a knot. Be sure the strings are all of the same length. Use a string about 4 in. (10 cm) long to attach the washer to the knot in the parachute strings. Make a larger parachute using a 24 in. (60 cm) square of plastic and the four remaining strings. Attach the washer to the parachute with a 4 in. (10 cm) piece of string as before. To test the parachutes, hold each in the centre of the plastic sheet. Flatten the plastic. Fold the plastic in half. Loosely wrap the string around the folded plastic.

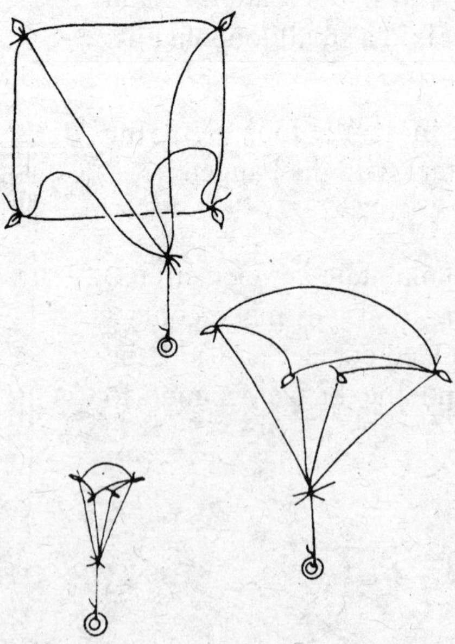

Throw the parachutes up into the air one at a time, and observe the time it takes for each to reach the ground.

This is What Happens

The larger parachute opens and floats to the ground more slowly than does the smaller parachute.

Science Behind It

Objects push against air as the force of gravity causes them to fall. This upward push of the air molecules is called air resistance. Objects with a large surface, such as the large parachute, have more air resistance. If the object has a large surface and a small weight, the upward push of air can equal the downward force due to weight, causing the object to gently float downward like a feather. Some insects have such a small weight as compared to their surface area that instead of falling from a height, they float down and therefore are not harmed.

90. Home Made Telephones Network

Imagine five empty match boxes positioned one right after the next. Push over the first match box and what happens? The first match box falls into the second. The second falls into the third. The third falls into the fourth and so on.

Imagine the tiny particles (atoms and molecules) that make up a kite string. In a similar way, motion can be transferred from particle to particle. Although you can't see this transfer, you can observe its effect using a string telephone.

Material Required

- Three paper cups;
- Kite string;
- Three paper clips;
- Pair of scissors;
- Pushpin;
- Some friends.

Procedure

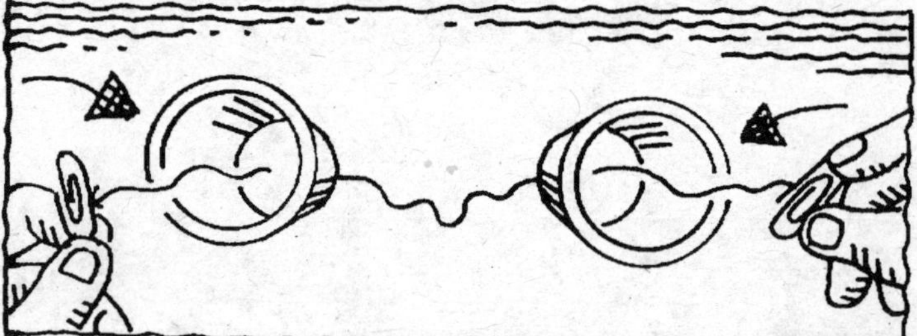

Use a pushpin to punch a hole in the bottom of each paper cup. Cut a segment of kite string about 20 feet long.

Pass one end of the string through the bottom hole of one cup. Tie a paper clip to the end of the string that projects from the inner side of the cup. Pull on the string so that the paper clip lies flat against the inside bottom of the cup.

Place a second cup and paper clip on the other end of this string. Move the two cups apart so that the line between them is pulled taut. Have a friend whisper into one cup. Place the other cup over your ear. Can you hear what your friend is saying? Reverse the roles. Can your friend detect your whispers?

Now cut a 10-foot segment of kite string. Place a cup and paper clip at one end of this line. Tie the free end to the middle of the two-cup line. Does this three-party line also transmit sound? How many more extensions can you add? Make a guess and then invite some friends to build a large string network.

This is What Happens

As one person spoke into the microphone cup, the movement of air struck the cup's solid bottom, causing it to vibrate. This energy of vibration was transferred into the string. Like falling match boxes, the atoms of the strings passed the vibrations from atom to atom. Eventually, the string vibrations reached the listening cups. At these cups, the string movements were transferred to the cup bottoms. The vibrating bottoms set the air within the cup in motion. As this vibrating air struck your eardrum, you detected it as sound.

91. The Magnetic Field

Electricity carrying cables are often buried underground. To locate their exact position, utility workers use a tool that detects magnetic fields. Within the cables, a current of flowing charges produces a magnetic field that extends into the surrounding space. By uncovering this magnetic field, workers can locate the exact position of the unseen electric flow.

Material Required

- Two "D" cells in battery holders;
- Connecting wires with each end stripped 1 1/2 inches bare of insulation;
- Switch;
- Index card;
- Iron filings;
- Compass.

Procedure

Assemble the complete circuit shown below. Place an index card on top of a section of wire so that you just can't see it.

Turn the switch to the "on" position for a few minutes only. Otherwise, you will damage the cell. Lightly sprinkle iron filings onto the card. Tap the card gently. Do the filings form a pattern or do they scatter randomly? Can you explain your observations?

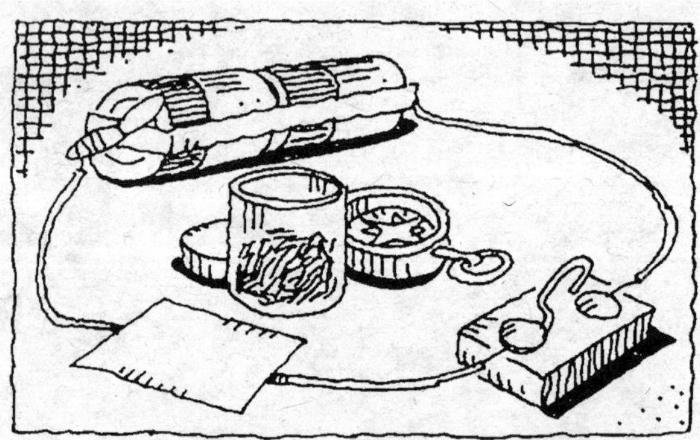

Turn the switch to the "off' position. Carefully return the iron filings to their container.

Place a compass alongside the wire. Note the direction that the needle points. While watching the needle, turn the switch to the “on" position. What do you observe? Explain what you see.

Suppose you switched the connectors to the cells so that their polarity was reversed. How would that affect the direction in which the needle points? Make a guess and then find out by switching the connections. Was your guess correct?

This is What Happens

Electric currents produce a detectable magnetic field. The filings that fell onto the index card "felt" the force of the magnetic field. Their final position formed a pattern that followed the magnetic lines of force that extended outward from the wire.

Likewise, the compass needle also "felt" the magnetic force. When the current flowed, it produced a magnetic attraction that successfully competed with the Earth's magnetic field.

92. Identical Speed

Does gravity pulls all things down at the same speed? Let's check it out in this experiment.

Material Required

- Paper;
- Book larger than the paper.

Procedure

Hold the paper in one hand and the book in the other. Drop the book and the paper at the same time. Observe the paper and book as they fall and strike the floor. Place the paper on top of the book. Do not have any of the paper hanging over the edges of the book. Hold the book waist high and drop it. Observe the paper and book as they fall and strike the floor.

This is What Happens

The book hits the floor before the paper when they are dropped separately. When the paper is on top of the book, it falls with the book, and both—book and paper—land at the same time.

Science Behind It

When the paper is lying on the book, the objects fall together because gravity equally pulls on both objects. Air pushing up against an object slows the speed of the falling object. The book's speed is not changed much because its weight (downward force) overcomes the upward force of the air. The paper's weight equals the upward force of the air: therefore, it falls at a slower speed.

93. Hanging Sounds

A hanger is good to hang cloths on it, right! Wrong, you can also build an incredible sound-making device from a hanger? Here's how.

Material Required

- Metal clothes hanger;
- Kite string;
- Metal spoon;
- Two paper clips;
- Two paper cups;
- Pair of scissors;
- Pushpin.

Procedure

Use a pushpin to punch a hole into the bottom of a paper cup. Cut two segments of kite string, each about 2 feet long. Pass one of the segments through the cup hole. Tie the end of this string segment that exits the inner side of the cup to a paper clip.

Repeat the above steps with another cup. When you're finished, you should have two listening devices that look as if they were cut from a string telephone.

Tie the free end of both strings to the hook of a metal hanger. Place each cup over your outer ear. Make sure that the hanger dangles freely.

Have a friend gently tap the hanger with a metal spoon or similar object. What do you hear?

This is What Happens

Sound travels very well through solid materials. When the hanger was struck by the spoon, it vibrated. These vibrations were transferred to the string. The string carried these back-and-forth movements to the cups. There, the vibrating cup bottoms set the air space in motion.

As particles of air vibrated within the cup, they struck your eardrum. You detected this movement as sound. Since the sound travelled along the solid string, it contained much of its original energy and produced a noise that sounded like the ringing of a huge bell.

94. Flipping Switches

Can two switches control a light from two different locations? Find out in this experiment.

Material Required

- Two "D" cells in battery holders;
- One flashlight bulb in lamp holder;
- Six thumbtacks;
- Two metal paper clips;
- Two pieces of wood or thick foam core;
- Connecting wires with each end stripped 1½ inches bare of insulation.

Procedure

For this activity, you'll need to build two new switches. The switch bases are made from two small blocks of wood. Carefully push three thumbtacks partially into the wood base as shown below. Bend apart a paper clip and wrap it around the thumbtack that is on the side by itself. Don't push the tacks fully into the wood until the wire connections are in place.

Now assemble the circuit below using these two switches. Unlike the previous switches, these circuit parts are sliding switches. The clip is constantly pushed against the top of the tack or wood surface. When you slide the bent clip onto a tack, you complete a circuit. This circuit remains "on" until you slide the clip off, the tack.

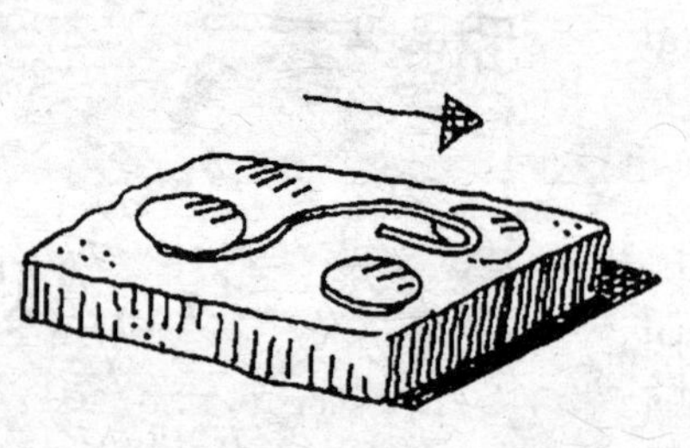

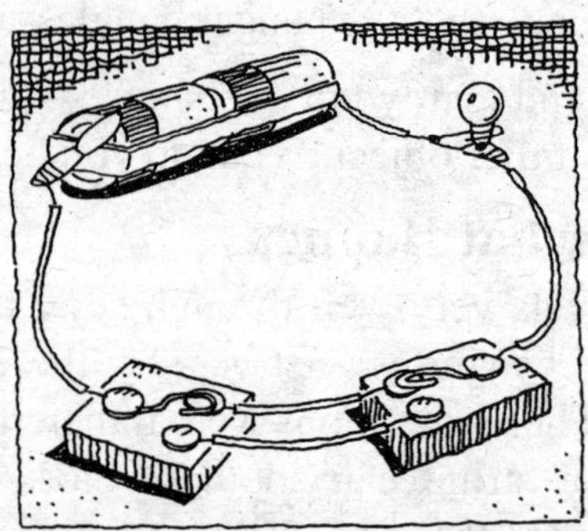

Flip one switch. What happens? Flip the other switch. What happens now? Which switch controls the lamp? Explain.

This is What Happens

The circuit is constructed so that both switches form the path over which the electric current must flow. If either switch is flipped, the circuit will change. If it was open, a flipped switch will close the circuit. If it was closed, a flipped switch will open the circuit. As you might infer, this type of switch is useful when you want to control a light from two different locations.

95. Pendulum Puzzle

Try this exciting experiment in playground with swing and find out if weight has any effect on the speed of a pendulum's swing.

Material Required

- Watch;
- Outside playground swing;
- Ruler;
- 2 helpers.

Procedure

Hold the seat of swing and move back 3 or 4 steps. Have your helper place the ruler on the ground in front of your feet. Ask your helper to start timing when you release the swing and to report the end of 10 seconds.

Caution: Do not push the swing, just release it.

Count the number of swings back and forth in 10 seconds. Ask one of your helpers to sit in the swing. Pull the swing back until your feet are behind the ruler as before. Again have your helper start the timer when you release the swing and report the end of 10 seconds.

Caution: Do not push the swing.

Count the number of back-and-forth swings in 10 seconds.

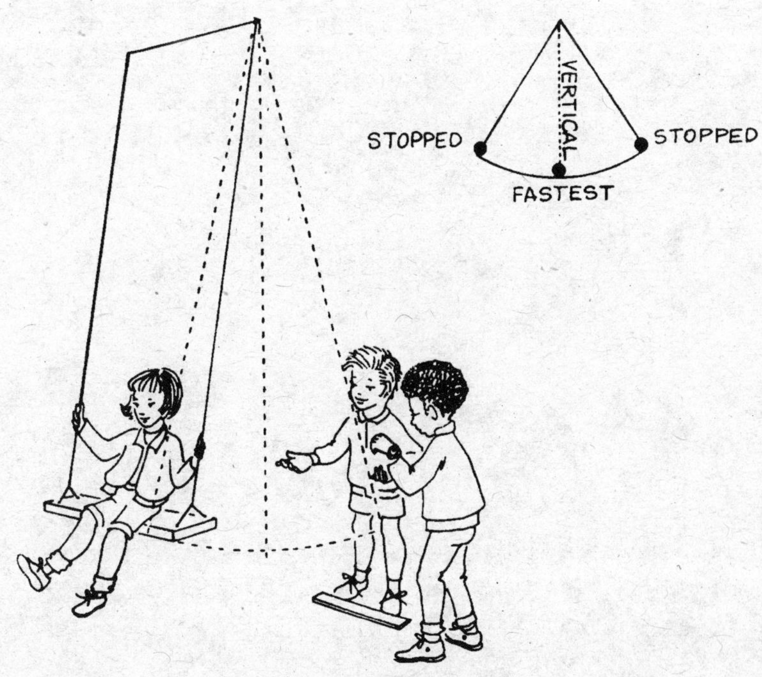

This is What Happens

The number of back-and-forth swings is the same with and without a person sitting in the swing.

Science Behind It

The number of back-and-forth swings was the same regardless of the weight placed on the swing because gravity pulled on the swing, causing it to fall when released. The speed during the swings changed, but the changes were the same for each weight that was swung. The speed increased as the swing or pendulum approached the vertical position and slowed as it moved upward where it stopped. Pendulums stop at the highest position of their swing before beginning the downward swing (that is due to the pull of gravity). The speed of this back-and-forth movement stays the same for each weight because the beginning height of the pendulum's swinging path remained the same each time and the downward pull on gravity is the same on all substances regardless of their weight.

96. Singing Balloon

Place your hand gently over your throat. Hum a tune. What do you feel? Can you detect any changes in the vibration as you vary the pitch and volume of the notes? Within your throat is a body part called a voice box. The voice box contains bands of muscle tissue called vocal cords. Like other muscles, the vocal cords can relax or tighten up. As air passes over these muscles, their tension "shapes" the sound that you create.

Material Required

- Balloon.

Procedure

Blow up a balloon and hold the open end shut. While preventing air from escaping, pinch both sides of this nozzle. Now slowly loosen your hold of the balloon's opening to allow a controlled flow of air through. Then pull the opening wider. What happens to the pitch?

Release the "stretch" so that the nozzle material relaxes. How does this affect the pitch?

This is What Happens

The nozzle was a model of your vocal cords. Air that passed through the nozzle opening caused the elastic material to vibrate. This vibration produced sound.

When the nozzle was stretched, the vibrating material became tighter. This change produced a note of higher pitch.

When the nozzle material was relaxed, the pitch dropped. By altering the tension in the nozzle, you produced a sliding scale of notes.

97. Simple Multimeter/Current meter

An electrician usually uses a special tool to measure the current that flows through a wiring. The tool is called a multimeter and it is capable of measuring several properties of electric flow. In this expert invent, you'll get to build a simple type of meter that will also respond to the flow of current.

Material Required

- 10 yards of 24 -gauge insulated copper wire;
- Wire strippers;
- Packaging tape;
- Sewing needle;
- Thread;
- Strong bar magnet;
- Juice can;
- "D" cell battery.

Procedure

Strip several inches of insulation from both ends of the wire. Wrap this wire around the can into a tight coil, leaving about a foot of wire on each end of the coil. Slip the coil off the can. Wrap several pieces packaging tape around the coil to secure its shape. Use a larger piece of tape to tape the coil onto a flat surface so that it stands up.

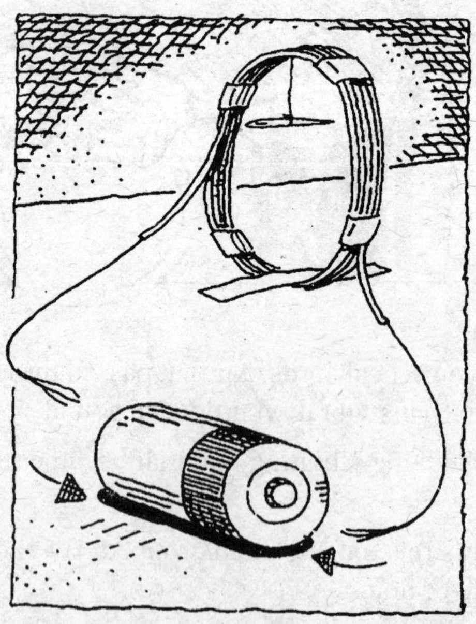

Use the magnet to magnetize the sewing needle. Stroke the needle at least forty times in the same direction.

Tie a small length of thread around the middle of the needle so that it balances. Tape the other end of the thread to the top of the wire coil, the magnetized needle should balance in the middle of the coil's open space.

Touch the free ends of the wires to the positive and negative sides or the "D" cell battery. What happens? Switch the wires and touch the opposite sides of the cell. What happens now?

This is What Happens

When the wires were attached to the cell, electricity flowed through the coil. This movement of charge created a magnetic field. The magnetic field affected the magnetized needle, causing it to spin and change its pointing direction. When the wires were switched, the current flowed in the opposite direction. In response, the needle turned and pointed in the opposite direction.

98. Time Effect

The idea of this labour is to determine how the length of a pendulum affects the time of each swing.

Material Required

- String;
- Washer;
- Scissors;
- Ruler;
- Heavy book;
- Table;
- Stop watch or watch with a second hand;
- Helper.

Procedure

Cut a string the height of the table. Tie one end of the string to the washer, and use tape to attach the other end of the string to the end of the ruler. Lay the ruler on the table with about 4 in. (10 cm) of the string extending over the edge, and the string hanging freely. Lay the book on top of the ruler to hold it in place. Pull the washer to one side and release it.

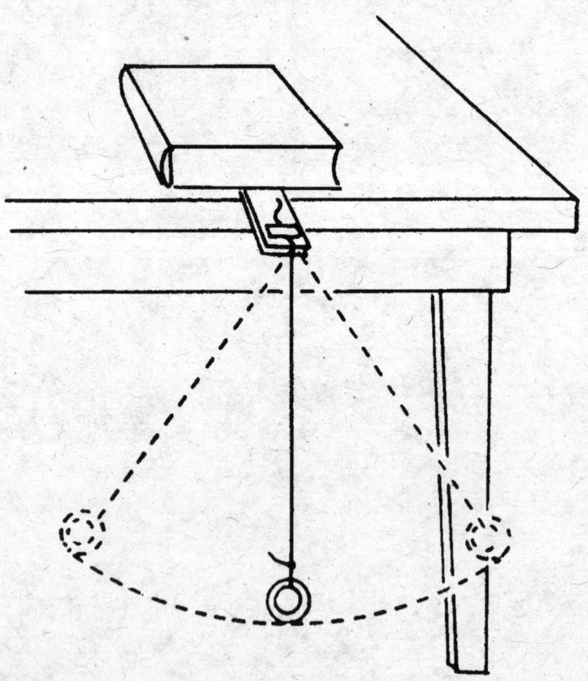

Ask your helper to start the timer as you count the number of swings in 10 seconds. Shorten the string by one-fourth its length. Pull the washer to one side, release it, and count the number of swings in 10 seconds as your helper records the time.

This is What Happens

The number of swings doubles with the shorter string.

Science Behind It

Galileo has been credited for discovering the relationship between the length of a pendulum and the time of its swing. The story told is that he observed the swinging of a great lamp while in church and timed the swings by comparing it with his pulse beat. He later discovered that the time of a swing depends on the length of the pendulum and that the time decreases by one-half if the string is one-fourth the original length.

99. Making Kazoos

Kazoo? I am not talking about the dry fruit, Kazoo. Kazoos are great fun, especially for those who don't have a great singing voice. Just hum into the mouthpiece and the vibrating drum enriches your sound with a chorus of buzzing plastic. Although you can purchase kazoos at local toy stores, here's one that you can construct at home.

Material Required

- Clean cardboard tube (removed from paper toweling);
- Pair of scissors;
- Cellophane-type plastic;
- Tape.

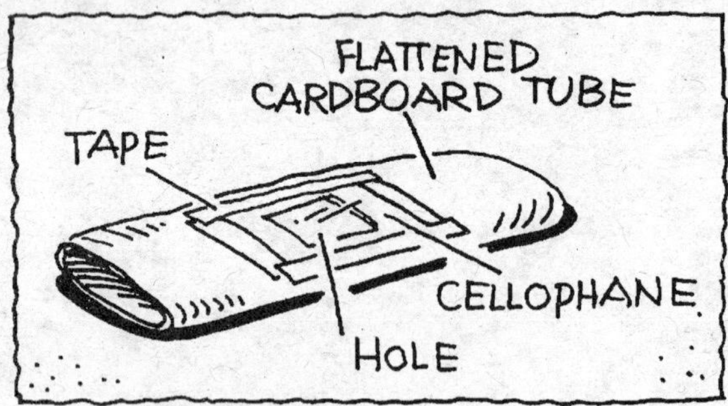

Procedure

Find a scrap piece of cellophane-type plastic. This clear, light-gauge plastic is easily recognized by the "crinkling sound" it makes when crumpled. Cut off a section of cardboard tube that is about 6 inches long. Squish down this tube.

Use your scissors to carefully cut a rectangle about the size of a postage stamp near the middle of the tube. Cover the hole with cellophane. Use tape to secure all four edges of the plastic to the tube. Hum into one end of the tube. What do you hear? How is the sound produced? Can you change the characteristics of the sound by using different-sized holes? Try it and find out!

This is What Happens

As you hummed into the tube, you created sound waves. These waves travelled along the inside of the tube. As they passed the cellophane, the sound waves energized the plastic, causing it to vibrate. The vibrations of the plastic generated sound waves. This buzzing noise added to the hum and produced a richer, fuller, and "buzzier" sound.

100. Lemon Cells

In this experiment, you will make current from ordinary lemons.

Material Required

- Lemon;
- Copper coin;
- Strip of zinc (obtained from hardware store);
- Steel wool;
- Knife;
- Current meter (assembled in "Simply Multimeter").

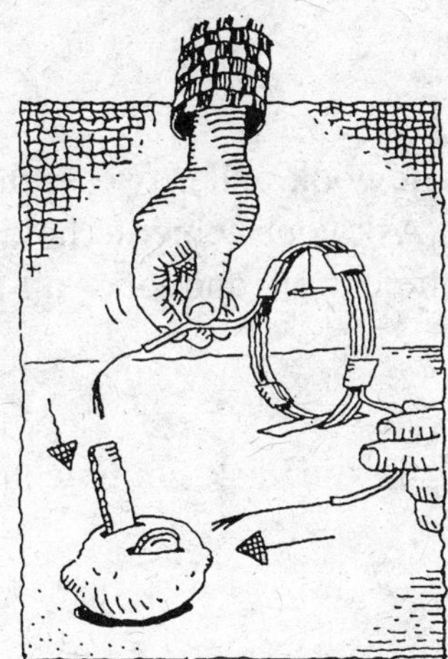

Procedure

Use the steel wool to shine the surfaces of the coin and the zinc strip. File down any sharp edges of the zinc strip.

Have an adult use the knife to punch two small slits into the lemon's tough skin. The slits should be about ½ inch in length and placed about ½ inch apart.

Insert a coin into one of the slits. Insert the zinc strip into the other slit. Make sure that the metals don't touch.

Touch the leads of the current meter to the exposed metals. What happens to the magnetized needle? Can you explain your observation?

This is What Happens

There is a natural tendency for electric charges to travel between two different metals (in this case copper and zinc). Within the lemon, the acid environment offered a partial route for the travel of charges. The route was completed by the external circuit, which included the coiled wire. As the current travelled through this coil, it produced a magnetic field. This field deflected the needle from its original pointing direction.

101. Need for Speed

How shape affects speed? In this, we are going to find an easy answer to the above question.

Material Required

- Card table;
- 2 books of equal thickness;
- 1 large roll of masking tape;
- 2 jar lids, the same size;
- Marble;
- Masking table;
- Helper.

Procedure

Tilt the card table by placing one book under two of the legs. Place the lid tops together and tape their edges to form a disk. Ask your helper to hold the disk made from the lids at the top of the incline while you hold the marble and tape roll in line with the disk. Release all three objects at once.

This is What Happens

The marble rolls fastest, with the lid disk coming in second, and the tape roll last.

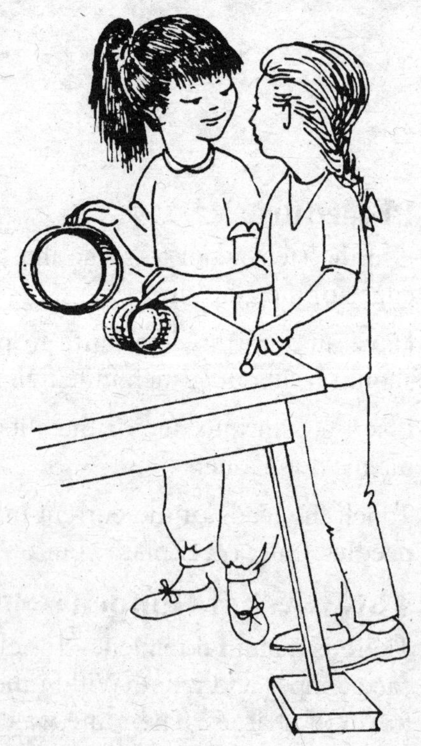

Science Behind It

The rolling speed is related to the distribution of weight around the object's centre of gravity (the point where the weight appears to be located). The closer the weight is to the centre of gravity, the faster the object can rotate. The centre of gravity of all the objects in this experiment is at their geometric centre, but each has a different weight distribution. The hollow tape roll's weight is located farthest away from the centre of gravity. It has the slowest rotation speed. The marble's weight is closest to its centre of gravity, resulting in it having the fastest rotating speed.

102. Musical Bottles

Like clarinets and saxophones, flutes are woodwind instruments. The flute's column of air, however, is not set in motion by a vibrating reed. Instead, the flutist must blow across the opening of the mouthpiece. When the stream of air flows through this hole at the right angle, it produces a soft "whistle." As this whistle travels through the instrument's body, it is shaped into the rich, warm tones of a flute.

Material Required

- Several plastic soft drink containers;
- Water;
- Ruler.

Procedure

Clean and dry an empty soft drink container. Place the neck of the container just below your lower lip. Curl your upper lip outward so that it directs a stream of air at the neck's opening. Vary the force of the air jet (and its angle) until a rich sound is produced by the flowing air.

Fill a second bottle about half full of water. Before you produce a sound with this container, make a guess. Will this note be higher or lower in pitch than the note produced in the empty container? Test your prediction.

Magic Words

If you have three identical containers and three friends willing to join in, you can form a pleasing musical chord. Keep one container empty. Fill another container one-fifth full with water. Fill the third container one-third full of water. Play the three notes at the same time. You may have to "tune" the containers slightly. This can be done by adding or removing water.

This is What Happens

As you blew into the bottle, you set the air in motion. As the particles of air moved back and forth, they formed a sound wave. This sound emerged from the neck of the bottle as a distinct note.

Water that was added to the container filled up some of the space. Since the column of air was shortened by the rising level of water, it produced a sound of higher pitch.

103. Battery of Cells

A single cell made of two metals can produce only a small "push" of charge. To increase this push, several cells can be wired together. This arrangement of side-by-side cells forms an electrical device known to scientists as a battery.

Material Required

- Three copper coins;
- Three iron washers;
- Steel wool;
- Blotter paper;
- Pair of scissors;
- Salt;
- Water;
- Current meter (assembled in "Simple Multimeter").

Procedure

Polish the surface of the coins and washers with steel wool. Use a pair of scissors to cut the blotter paper into four coin-sized circles.

Soak the paper in salt water. Sandwich a damp paper circle between a coin and a washer. Test the generation of electricity with your current meter. How does the meter react?

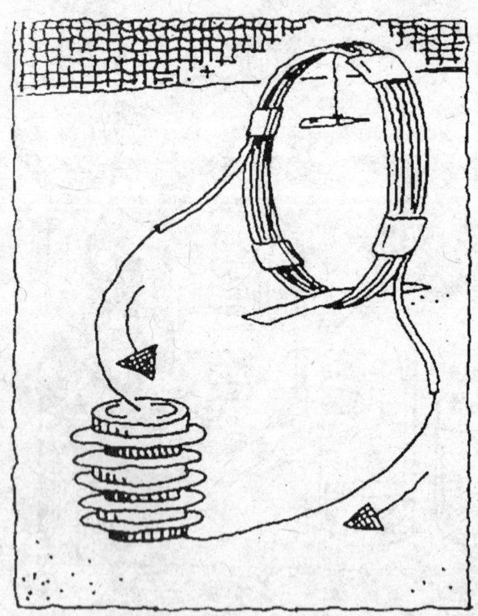

Make two other sandwiches. Stack all three sandwiches into a pile of alternating coin/wet paper/washer units. Insert a saltwater-soaked paper between the coin of one cell and the washer of its neighbour. In other words, no two coins should make direct contact.

Use tape to secure this three-cell stack. Retest the current. Is it stronger? Can you explain your observations?

This is What Happens

The simple coin/washer cell produced a very small amount of electricity. As the cells were joined together, they formed a battery. The three-cell battery generated three times the electricity of a single cell. This increase produced a more noticeable deflection of the meter's needle.

104. Hanging Hammer

This experiment is almost magical in nature. The purpose is to demonstrate that the centre of gravity is the balancing point of an object.

Material Required

- Yardstick (meter stick);
- Hammer (wooden handled hammer works best);
- String, 12 in. (30 cm).

Procedure

Hold the ends of the string together and tie a knot about 2 in. (5 cm) from the ends. Insert the hammer and yardstick through the loop. Position the end of the yardstick on a table's edge. The handle of the hammer must touch the yardstick and the head of the hammer will extend under the table. Change the position of the hammer until the whole unit—yardstick, string, and hammer—balances.

This is What Happens

The unit balances with only a small amount of the yardstick touching the table.

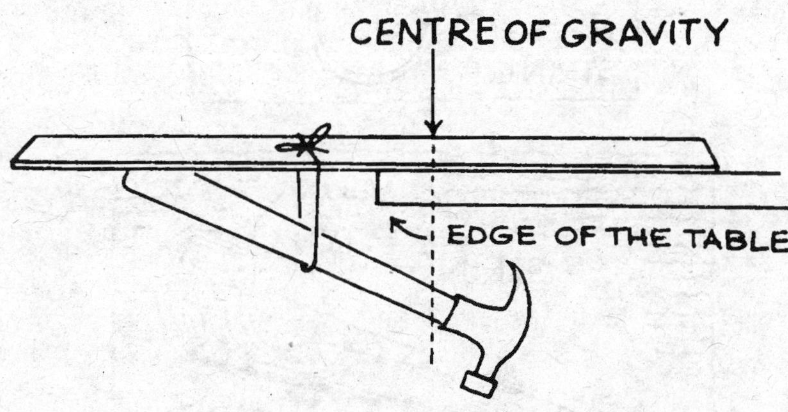

Science Behind It

The hammer, string, and yardstick all act as a single unit with a centre of gravity. The centre of gravity is the point where any object balances. The dashed line in the diagram allows you to visualize the centre of gravity. The heavy hammer head counterbalances the weight on the left side of the balancing point.

105. Pitch Switch/ Making Ektara

Slide your fingers along a guitar neck and you'll feel ridges. These ridges divide the neck into distinct regions called frets. When the string is pressed, it's vibrating length is determined by the nearest fret. Frets closest to the sound hole produce the highest notes. Frets farthest from the sound hole produce notes of lowest pitch.

Material Required

- Paper cup;
- Kite string;
- Tape;
- Ruler;
- Two paper clips;
- Pushpin.

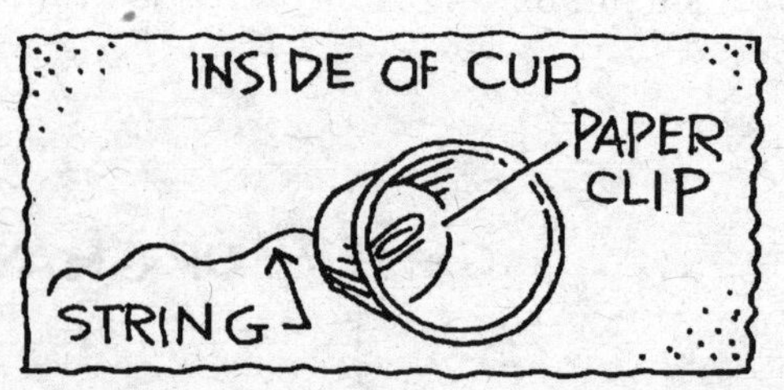

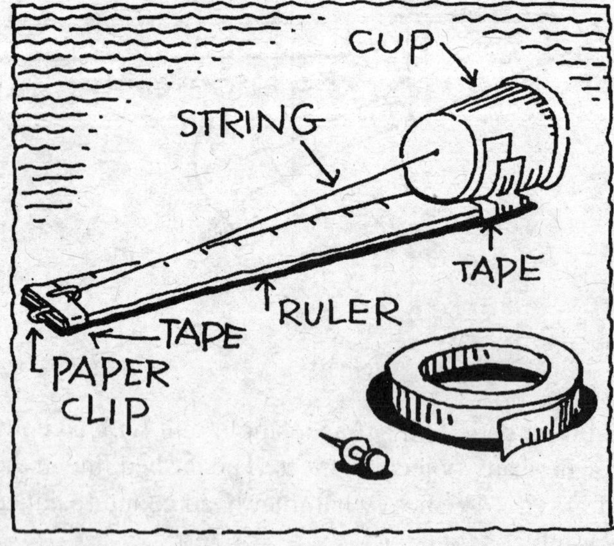

Procedure

Use the pushpin to punch a small hole near the centre of the cup's bottom. Cut a length of kite string about 2 feet long. Pass the string halfway through the cup hole. Tie the end of the string (on the inner cup side) to the paper clip.

Pull the opposite end of the string so that the paper clip becomes pressed against the inside of the cup bottom. Use tape to secure the cup to one end of the ruler.

Bend the other paper clip into an "S" shape hook. Fasten this hook to the opposite end of the ruler. Use tape to secure.

Tie the free end of the string to the paper clip. The string should be tied so that it is taut and can be "plucked."

Trim away any excess string. Pluck the string. Describe the sound. Press down on the string to force it to contact the ruler. Pluck the string again. What has happened to the sound? Try pressing down at different points along the string. How does the location of the "press" affect the sound?

This is What Happens

When the string was first plucked, it vibrated freely along its entire length. This produced the sound with the lowest pitch. When you pressed down on the string, you shortened the length of the vibrating section. This shorter length produced a higher pitch. By moving this "end point," you varied the pitch of the plucked note.

106. Generating Electricity

So far, the electricity you've generated has been produced by chemical reactions. Although this source of electricity is important, most of your home is supplied by electricity that has been produced by generators. Generators are large devices that change the energy of spinning magnetic fields into electric current.

Material Required

- Strong bar magnet;
- 10 yards of 24-gauge insulated copper urine;
- Juice can;
- Tape;
- Current meter (assembled in "Simple Multimeter").

Procedure

Strip several inches of insulation from both ends of the wire. Wrap this wire around the can into a tight coil. Leave about a foot of wire at each end of this coil. Slip the coil off the can and secure its shape with several pieces of tape. Attach the bare ends of these wires to the ends of the current meter's wires. A complete circuit containing two coils should be formed.

Push the bar magnet in and out of the newly assembled coil. What happens to the needle that is suspended in the current meter? Can you explain your observations?

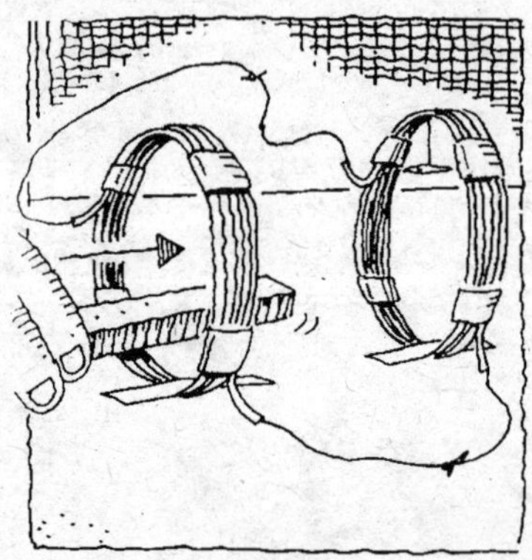

This is What Happens

You've just generated electricity. As the magnet moved in and out of the coil, it induced a back-and-forth flow of charges within the coil. This flow moved throughout the circuit. At the meter, this alternating current produced a flip-flopping magnetic field. Its effects were seen by the needle, which continually changed directions.

107. Straw See-saw

You may have played on see-saw in the parks or gardens. But how does centre of gravity affects a balance? This is what we are going to find out.

Material Required

- Straight pin;
- Small index card;
- Straw;
- Scissors;
- 2 wooden blocks of equal height and not as wide as the length of the straw;
- Ruler;
- Marking pen.

Procedure

Use the ruler to find the centre of the straw and mark the spot with the marking pen. Cut a 1 in. (2.5 cm) slit in each end of the straw, as shown in the diagram. The slits should be in the same relative position on each end. Divide the index card in half by folding and cutting along the fold. Insert the paper pieces in the slits on each end of the straw to form two flat surfaces that are parallel with each other. These papers will act as weighing pans.

Punch the straight pin through the centre of the straw, leaving an equal amount of the pin sticking out on each side of the straw. Position the two wooden blocks on a table and place the ends of the pin on the edges of the blocks. Move the paper pieces outward to increase the downward pressure on the weighing pan and inward to decrease the pressure. Position the paper weighing pans to make the straw level with the table.

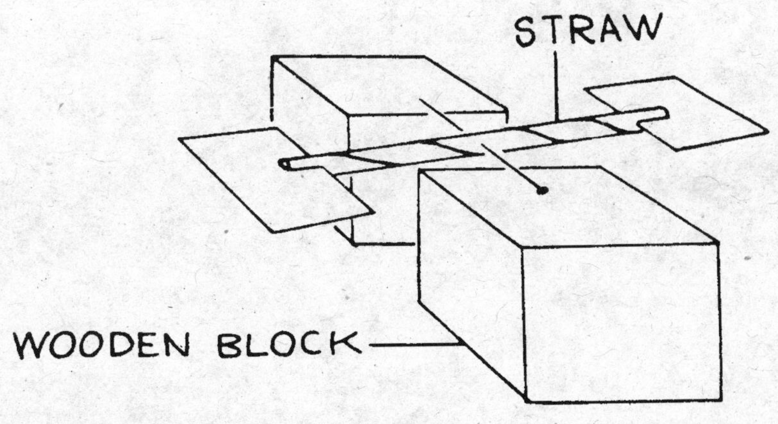

This is What Happens

Moving the paper weighing pans in and out causes the straw to rotate around the pin. A position was found to make the straw level with the table.

Science Behind It

The position of the paper weighing pans affects the balance of the straw. The farther the paper is moved away from the pin, the more downward is the rotation of the straw on that side. Children of different weights experience the same thing when sitting on a see-saw. They move away from the centre of the board to increase their downward rotation and closer to the centre to decrease it. The straw is balanced when the position of the paper places the centre of gravity of the balance at the place where the pin is inserted. Centre of gravity is the point where an object balances. The downward rotation on both sides of the centre of gravity point is always the same.

108. Sound Frequency

You might have seen in Orchestra. Musicians need to stay in tune. That's where a tuning fork comes in. A tuning fork is a piece of metal that, when struck, vibrates at a very exact rate. This rate of vibration is called a frequency. Some tuning forks vibrate at the frequency of a C note. Other forks vibrate at the frequencies of other notes. By selecting the correct tuning fork, a musician can tune an instrument to the same pitch that is shared by all band members.

Material Required

- Tuning fork;
- Cup filled with water.

Magic Words

When striking a tuning fork, hit it against a solid but padded object. Knees, elbows, cloth-covered furniture, or carpeted floors work great. Try not to hit the fork against hard, unpadded objects. A hard strike might bend or distort the shape of the tuning fork. If this happens, the tuning fork may change its vibration and produce a note of different pitch or less volume.

Procedure

Firmly hold the tuning fork by its handle. Strike one of the double ends against your knee or padded surface. Keep holding the handle. Do you hear the note? Strike the fork again. Is the note the same or different?

Fill a cup with water. Strike the tuning fork against your knee. Slowly bring the vibrating ends to the water's surface. Touch these ends to the water's surface. What do you observe?

109. Morse Code Transmitters

With the help of this experiment, you can construct your own Morse Code Station.

Material Required

- Four "D " cells in battery holders;
- Two flashlight bulbs in lamp holders;
- Two switches;
- Four brass fasteners;
- Plenty of connecting wire with ends stripped bare of insulation.

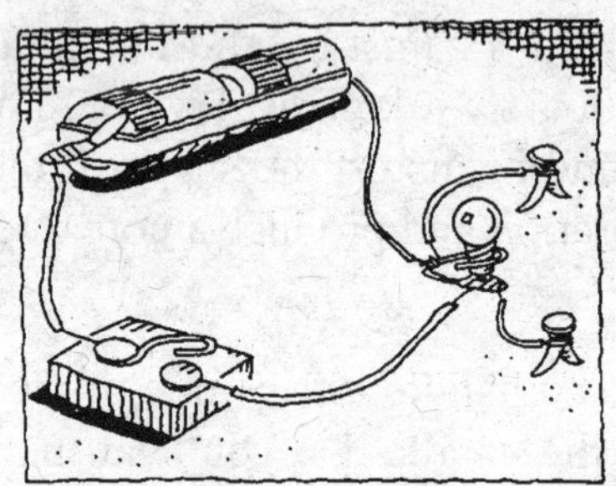

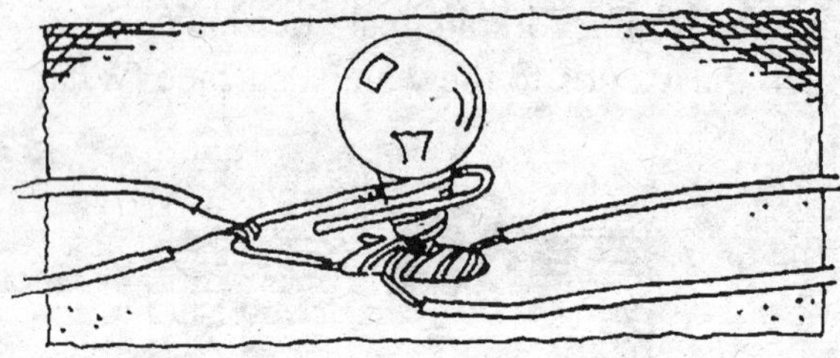

Procedure

Assemble two separate Morse code stations that look like this:

Next, place the two stations at different locations and connect them by wire. Begin by setting them up side-by-side. This will make solving any sort of connection problem much easier and quicker.

Use two lengths of connecting wire to attach the stations as shown here.

If this short connection works, you're ready to expand. Use more connecting wire and place the stations at opposite ends of your room. Do the transmitters still work? If so, you're ready.

Have an adult help you set up a pair of wires that will connect two stations that are placed in different rooms of the house.

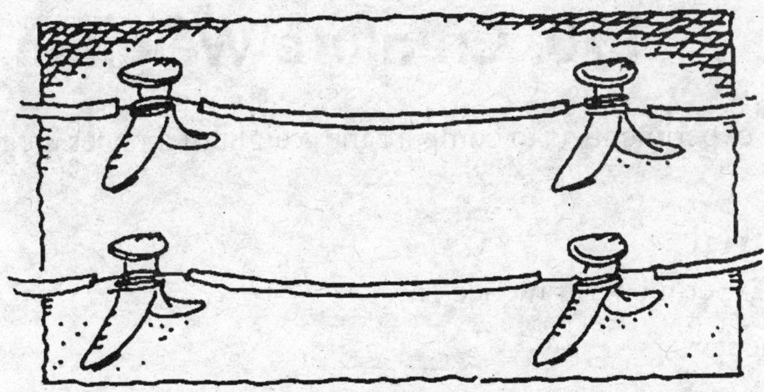

The Morse Code

Make two photocopies of the Morse code sheet below. Keep a copy by each station.

A B C D E F G

H I J K L M N

O P Q R S T U

V W X Y Z ERROR

1 2 3 4 5

6 7 8 9 0

Below each letter and number is the code that represents it. The circles are "dots." A dot is transmitted by a quick ON/OFF press or the switch. The small rectangles are "dashes." A dash is transmitted by a longer ON/OFF press.

This code was developed by a man named Samuel Morse. In 1838, Samuel invented the first device that could send coded messages along electric wires. Soon everyone was calling these wire messages Morse code.

This is What Happens

When either switch (called a key by Morse code people) is closed, the double-lamp circuit becomes complete. The current flows from the closest pair of cells through the large connecting loop of wire (that is strung between the stations). Flowing along this wire, the electricity lights both lamps. By monitoring your own bulb, you can make sure that your message has been sent to the other station. When you see a light, they will see a light.

110. Creature Weigh-in

The purpose of this experiments is to compare the weight of a paper creature with that of paper punches.

Material Required

- Balance from Experiment Straw See-saw;
- Paper hole puncher;
- Index card;
- Scissors;
- Pencil.

Procedure

Draw your version of a space creature on half of the index card. Cut out the creature and place it on one of the balance's paper weighing pans.

Cut paper punches from the remaining portion of the index card and continue to place them on the empty paper weighing until the straw is level with the table.

This is What Happens

The end holding the paper creature falls down, but starts to rise as paper punches are added to the opposite weighing pan. Too many punches lift the creature above the balancing point.

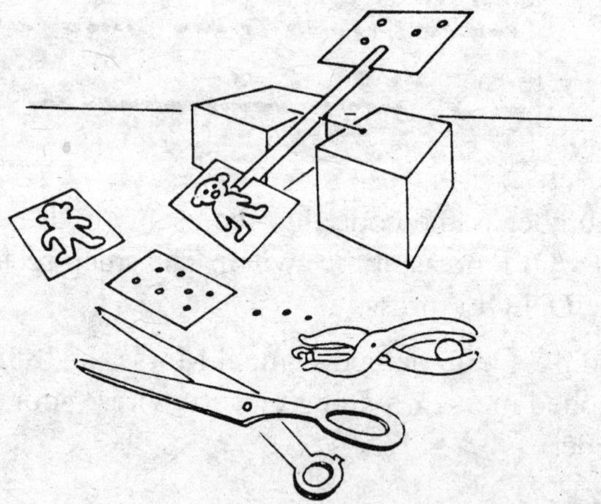

Science Behind It

The downward pull that gravity has on an object is called its weight. Placing the paper creature on one side of the balance increases the weight on that side. Adding the paper punches on the opposite pan begins to balance the weight of the creature. When the total weight of the paper punches equals the weight of the paper creature, the balance will be level with the table. The level balance indicates that the pull of gravity is the same on both sides of the balance.

111. Size and Sound

Look at the stage of a rock 'n' roll concert and you're likely to see a wall of speakers. Get close to these speakers and you'll discover that they are big, really big. Big speakers have a larger vibrating surface than smaller speakers. This larger surface area can transfer vibrations to many more air particles than a smaller surface area. As the numbers of these vibrating particles increase, so does the volume!

Material Required

- Tuning fork;
- Large flat surface (such as a tabletop);
- A friend.

Procedure

Strike a tuning fork against your knee. Hold it at arm's length. Can you hear the pitch it produces? Have a friend stand across the room. Can he or she also hear the sound of the vibrating fork?

Now strike the tuning fork again. As it vibrates, place the base of the handle firmly against a large flat object, such as a tabletop or chalkboard. What happens now?

Experiment with surfaces of different size and composition. What characteristics make a surface boost the loudness of a sound? What characteristics make an object muffle the loudness of a sound?

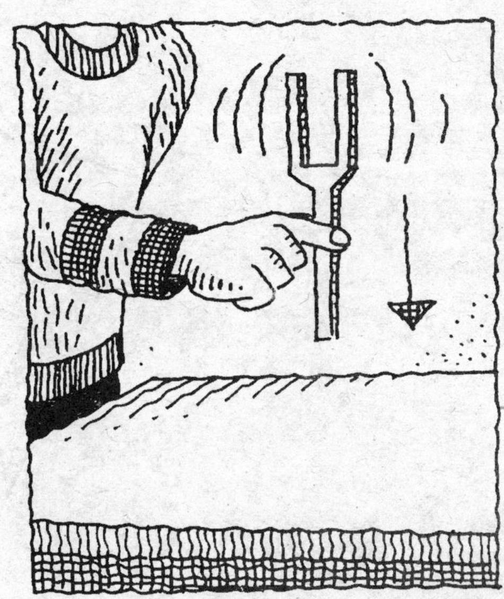

This is What Happens

As the tuning fork vibrates, it transfers its vibration energy to the surrounding air particles. As you can see, the arms of the tuning fork are quite small. Therefore, they are not in direct contact with a great deal of particles. Sounds produced solely by the vibrating fork aren't loud.

When the fork is placed against a large surface, such as a tabletop, things change. As the fork continues to vibrate, its vibration is transferred to the tabletop.

Because of its size, the tabletop is in contact with many more air particles than the tuning fork.

As these greater number of particles are set in motion, a much louder sound is produced. As you investigate materials, you'll find that hard surfaces that are free to vibrate are great for boosting sound volumes. In contrast, materials that cannot vibrate are poor boosters of sound.

112. Mapping Magnetic Fields

By using the compass, you will be able to map the invisible magnetic fields in this simple and easy experiment.

Material Required

- Bar magnet;
- Compass;
- Paper;
- Pencil.

Procedure

Place a magnet in the centre of a sheet of paper and carefully trace out its shape. Bar magnets are rectangular.

Now place the compass on the same sheet of paper a few inches from the magnet. Draw its outline onto the paper. Observe the direction that the compass needle points. Pickup the compass and draw the direction in which the needle was pointing. Place the compass in another nearby spot. Again, trace its outline and draw in the direction of the needle. Keep going until a pattern emerges.

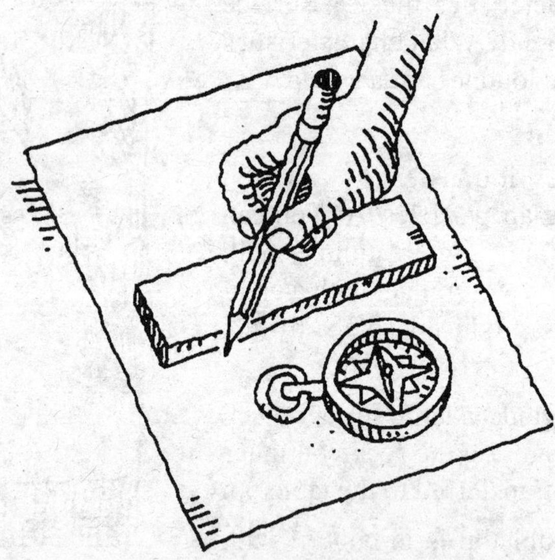

This is What Happens

A compass needle is a tiny, lightweight magnet. It reacts to both Earth's magnetic field and to the fields of any nearby magnets. When it was placed close to the bar magnet, the needle rotated so that it aligned with the invisible field of the magnet. By moving the compass, you were able to uncover the extent and direction of the bar magnet's lines of force.

113. Centre of Gravity

Where an object's centre of gravity is? This is what we are going to reach for in this experiment.

Material Required

- Paper hole punch;
- Push tack;
- Washer;
- String, 12 in. (30 cm);
- Scissors;
- Manila folder;
- Ruler;
- Tack board.

Procedure

Cut one side of the manila folder into an irregular shape. Punch four randomly spaced holes in the edge of the paper with the paper hole punch. Tie one end of the 12 in. (30 cm) string to the washer. Attach the free end of the string to the push tack. Stick the tack through one of the holes in the paper and into the tack board.

Allow the paper and string to swing freely. Use the ruler to mark a line on the paper next to the string. Move the tack to the other holes and mark the position of the hanging string each time. Do this for all four holes. Place the paper on the end of your index finger. Your finger is to be below the point where the lines cross.

This is What Happens

The paper balances on your finger.

Science Behind It

The earth's gravitational force—gravity—pulls downward on everything. Centre of gravity is the place on an object that acts as if all the gravitational force pulling on the object is pulling from that spot. The centre of gravity point of the paper is that point where the four lines cross. The paper can be balanced at this spot because all the paper's weight is evenly spaced around its centre of gravity. Hold your finger under that point and observe the balance.

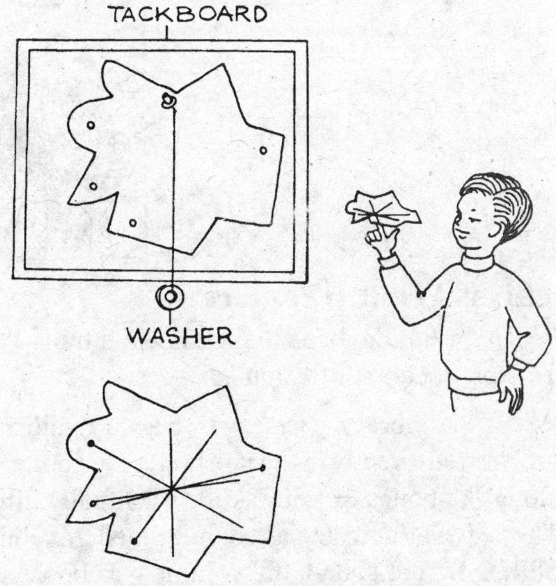

114. Listening Bones

Have you ever listened to a recording of your own voice? If so, were you surprised to hear how it sounded? Most people discover that their recorded voice sounds much different from the voice that they hear when they speak. The difference has to do with how the sound is conducted to your ears.

When you speak, much of your sound is projected outward into the air. A good portion of it, however, is captured by your body. This captured sound vibrates through your body and arrives at your ear through the bones of your head. This through-the-bone sound changes what you detect. Others (and tape recorders), however, will only hear the sound that is transmitted through the air.

Material Required

- Tuning fork.

Procedure

Strike the tuning fork against your knee. Listen to the sound. How would you describe its volume and richness?

Strike the tuning fork again. This time, while the fork is still vibrating, firmly press the base of the handle against the side of your jaw. How does pressing the fork against this bone affect the characteristics of the sound?

This is What Happens

When the tuning fork vibrated in air, it transferred its energy to a limited number of air particles. This produced a soft sound.

When the vibrating fork was placed against your jaw, things changed. The vibrations were transferred directly to the jaw bone. This bone vibrated and passed the back-and-forth movement along the bones of your skull. Eventually, this "internal vibration" was transferred to the ear. There, it was detected as sound. Since travel through solid objects, such as bone, is a great way of transferring sound, the volume was boosted.

115. Magnetized Links

As in every horror movie, when a person gets bitten by a vampire, he or she will soon transform into one of the undead. People who are then bitten by this new vampire will also become vampires. It's a deadly chain. From bite to bite, victim to victim, the existence of vampires continues. This experiment shows such properties of magnet.

Material Required

- Bar magnet;
- Dish;
- A handful of small iron washers.

Procedure

Fill a dish with small iron washers. Gently lay the magnet onto this pile. Slowly pick up the magnet. What do you see? Do the washers form a hanging bridge? What part of the magnet do they stick to? Can you explain why the washers behave as they do?

This is What Happens

A magnetic force can travel "through" magnetic objects. The poles are the strongest regions of the magnet. The magnetic force "flows" from these poles to the washers that are in physical contact with the magnet. Those washers now become magnetized and they act as magnets. Any washers in contact with these magnetized washers also become magnetized so that washers in contact with these washers become magnetized. Finally, the magnetized links from both ends of the mag-net meet up and form a hanging chain.

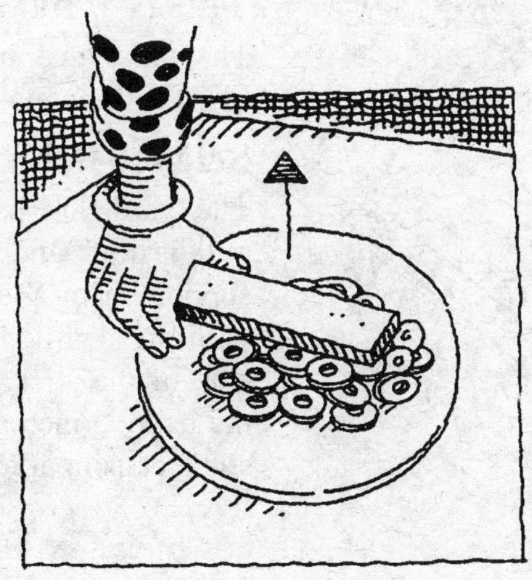

116. Weight of Air

Is air weightless or does it have any weight? This is what we are going to find out by performing this experiment.

Material Required

- Modeling clay;
- Pencil;
- Yardstick (meter stick);
- 3 balloons, 9 in. (23 cm); the balloons must be the same size;
- String, 4 ft. (120 cm);
- Scissors.

Procedure

Use the clay to secure the end of the pencil to the edge of the table. Cut four 12 in. (30 cm) strings. Suspend the yardstick by tying one of the strings to the centre of the stick and looping the free ends around the pencil. Adjust the position of the string in order to balance the yardstick. Use the precut string to suspend two uninflated balloons on equal distance from the centre support string. Move the balloons back and forth until the yardstick and the balloons balance. Inflate a balloon and attach a precut string. Make a loop with the string ends. Remove one of the uninflated balloons and replace it with the inflated balloon.

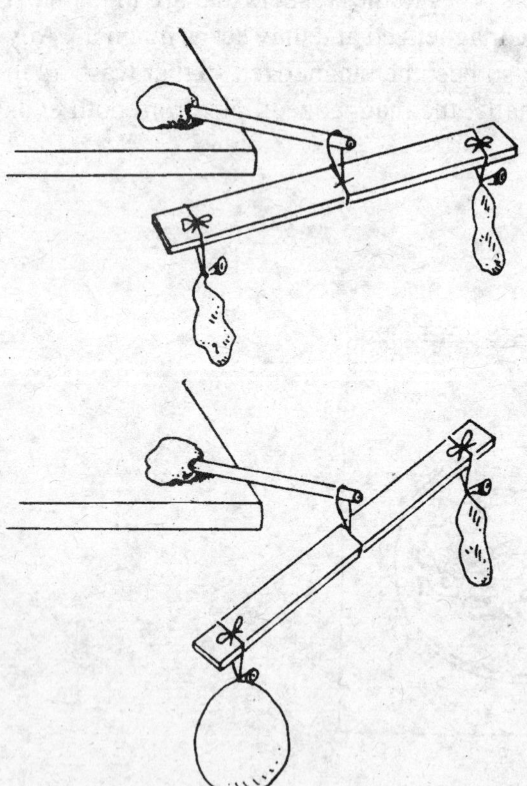

This is What Happens

The uninflated balloons balance, but the air-filled balloon makes the yardstick unbalanced.

Science Behind It

The yardstick balances when the downward pull is the same on both sides of the centre support string. Replacing the uninflated balloon with an air-filled balloon puts weight on one side of the stick. The weight of the air inside the balloon increases the downward pull on the stick, causing it to move down on that side.

117. Singing Bottles

Are you a bathroom singer? Have you ever sung in the shower? If so, did some notes sound much louder than others? By varying your pitch, you can uncover notes that the room seems to boost. This amplification is a natural effect that is caused by the shape of the bathroom space. When you hit the right note, the walls seem to join in and vibrate with your voice!

Material Required

- Two empty and identical soft drink containers;
- A friend.

Procedure

Place a soft drink container on a flat surface. Have a friend hold the other container several feet away. Instruct your friend to blow into the container to produce a low-pitched whistle.

After the whistle has sounded for several seconds, position your ear near the opening of the other container. What do you hear? How long does it last? Who started this tone?

This is What Happens

When your friend blew into the soft drink container, he or she produced a steady pitch. As this sound travelled outward from the "played" container, its waves struck the empty container. The empty container began to vibrate.

Since both containers were identical, they had the same built-in "pitch." When struck by the "tuned" sound waves, the unplayed container responded and began to vibrate. The vibration was strong enough to produce a back-and-forth movement of the air trapped within this unplayed bottle. Soon, both bottles were producing the same pitch.

118. Needles Pointing North

Two thousand years ago, Chinese navigators figured out how magnets could be used in navigation. Some people believe this connection was discovered when someone dropped a piece of natural magnetic rock (called lodestone) into a bucket of water. The rock settled so that it "pointed" to north. Every time it was dropped, the rock came to rest in the same direction. This is how compass was discovered.

Material Required

- Sewing needle;
- Magnet ;
- Plastic foam cup;
- Pair of scissors;
- A bowl half with water.

Procedure

Carefully remove the circular bottom of a plastic foam cup and place the bottom on the surface of the water. Observe how it floats.

Stroke the magnet along the side of a needle. Remember to stroke only in one direction. Place the magnetized needle on the floating foam bottom.

What happens to the needle and foam?

Now place the magnet along the edge of the bowl. Move the magnet around the bowl. Now what happens to the needle?

This is What Happens

Earth has a magnetic field. Magnetic objects placed within this field will react to this attractive force. When the magnetized needle was placed onto the floating foam, it spun freely. Under the influence of Earth's field, the needle (and float) rotated to its most stable position. Its north-seeking end pointed to the magnetic North Pole.

119. Balancing Performance

In this experiment, we will balance two forks on the edge of a drinking glass. Sound like magic trick? But my dear this is pure science.

Material Required

- 2 metal forks;
- Drinking glass or wide-mouthed jar;
- Modeling clay;
- 1 flat toothpick.

Procedure

Make a ball of clay about the size of a large marble. Insert the tip of one of the forks into the clay ball. Insert the second fork at about a 45-degree angle from the first fork. Insert the pointed end of the toothpick in the clay between the forks.

Place the end of the toothpick on the edge of the glass. Move it further over the glass until the forks balance.

Note: Decrease the angle between the forks if they will not balance.

This is What Happens

There is one point at which the toothpick supports the weight of both forks and the clay.

Science Behind It

The angle of the forks spreads their weight so that there is one place on the toothpick where all of the weight is evenly distributed. This spot is called the centre of gravity.

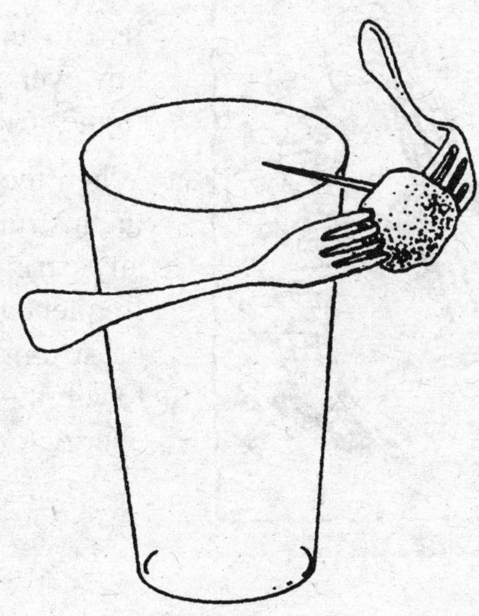

120. Ghost Sounds

Imagine being in a room where several people are playing musical instruments. In the corner, sits an unplayed piano. The music builds in intensity, filling the room with a loud and rich sound. Then the music ends abruptly. The piano, however, plays on. Although no one sits at the instrument, the piano produces a note that seems to fit in with the last note played by the musicians. Ghosts? No, resonance.

Material Required

- Guitar, piano, or other stringed instrument.

Procedure

Strum an open guitar string or play any key on the piano. Listen and memorize the pitch. You may have to sing with the note in order to produce its exact pitch. Practise until you can reproduce the same note.

Stop the instrument's string from vibrating. Once the string is stationary, sing out your note in a loud and steady tone for several seconds. Direct your voice at the sound box, sound board, or string itself.

Abruptly stop singing. Listen. What do you hear? Can you tell which string is vibrating? How can you be sure?

This is What Happens

By listening and reproducing the instrument's sound, you tuned your voice to the pitch of the vibrating string. When you touched the vibrating string, its movement was stopped. Your voice, however, produced a note that was "tuned" to the string.

When the vibrations of your voice struck the instrument's strings, they set in motion any strings that had the same built-in frequency. Even though it wasn't plucked or strummed, the original string started to vibrate. Once it started, it had enough energy to continue vibrating on its own.

121. Attractive Shapes

Water has a skin-like covering formed by the way water molecules attract each other. This attraction is called surface tension. Objects such as needles and pins float effortlessly on this invisible skin. Needles and pins are also easily magnetized. Oh? Perhaps we can combine these characteristics?

Material Required

- A dozen sewing needles or pins;
- Magnet;
- Non-metal bowl;
- Water;
- Metal paper clip.

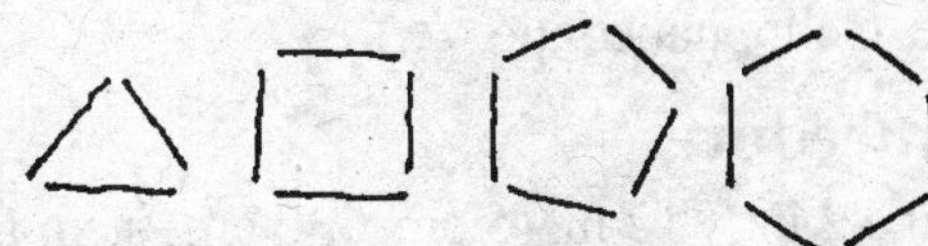

Procedure

Magnetize about a dozen needles by stroking a permanent magnet along their length. Remember to stroke in only one direction.

Fill a bowl three-fourths full with water. Put a bend in a paper clip, forming a right angle between its two loops. Use this paper clip as a "cradle" to lower needles onto the surface of the water.

Place several needles on the water's surface. Observe how their ends react to each other. What causes some of the needles to attract and others to repel?

Take a look at the shapes shown here. Can you steer the needles into the patterns below? Good luck.

This is What Happens

There are two major concepts at work here. The first is surface tension, which is formed from attractive forces between neighbouring water molecules. The second concept is magnetism. As you have learned, like poles repel, while unlike poles attract. In order for needles to align end-to-end, the adjoining ends must have opposite charges. Once they are positioned correctly, the magnetic force helps keep their shape.

122. Kite's Tail Tales

When flying kites, you may have seen tails behind them. Have you ever wondered what's the purpose of this tail is in the kite? In this experiment, we will find the effect of a kite's tail.

Material Required

- 1 sheet of notebook paper;
- Scissors;
- String;
- Ruler;
- Cellophane tape.

Procedure

Cut a 2 in. x 12 in. (5 cm x 30 cm) strip from the sheet of paper. Use tape to attach an 18 in. (45 cm) length of string to one end of the strip. Hold the free end of the string and whip the paper back and forth in front of you.

Cut a 1/4 in. x 12 in. (0.5 cm x 30 cm) strip from the paper and attach it with tape to the free end of the wider strip. Again move the strip back and forth in front of you.

This is What Happens

The paper twirls around, but when the small strip is attached the movement is smoother.

Science Behind It

The paper moves forward at an angle, causing the air to flow faster over the top side. Fast-moving air has a lower pressure around the moving stream. Thus, more uplift is exerted on the bottom of the strip. The angle of the paper is not constant, causing changes in the pressure along with a turbulent air flow across the strip. These changes make the strip twist and rotate. The paper tail makes the angle more constant. Therefore, there is a smoother flow of air across the paper and less twisting.

123. Stethoscope

A physician's stethoscope is a sound-capturing device. Its sound gathering end is formed by a flat drumhead. Sounds detected by this surface are transferred through the hose to the broadcasting end of the tool. There, vibrations leave the tube and are detected by the physician's ear.

Material Required

- Two funnels;
- 1-foot-long plastic tubing.

Procedure

Slip each end of the plastic tubing over the spout end of a funnel. To test your device, place one of the funnels against the wall. The rim of the funnel opening should lie flat on the wall's surface. Place the opening of the other funnel around your ear. Gently tap on the wall. What do you hear?

Now place one of the funnels over your heart. Make sure that the rim lies flat against your skin. Move the other funnel to your ear. What do you hear?

Keep listening. Take a deep breath. Can you hear the air as it travels along the passageways of your lungs?

This is What Happens

Your heart is a powerful muscle. As this muscle contracts, it pumps blood through the vessels of your body. To ensure that the pumped blood follows a one-way route, valves shut off and block the return flow of blood. The closure of these valves within the heart produce the sound we call a heartbeat. As this vibration travels through the body, it spreads out. The large collector of the stethoscope detects and concentrates these sounds. At the ear, the vibrations are detected as sound.

124. Floating Magnets or Anti-gravity Disks

Anti-gravity disks? You're joking.

Yes, we are. But the title probably did get your attention. These disks appear to float in mid-air. Their secret isn't an anti-gravity formula handed down by aliens from UFO's. It's just a result of magnetic repulsion.

Material Required

- Four magnets with holes in their centre (available at local electronics stores);
- 6-inch-long wooden dowel;
- Flat wooden base;
- Carpenter's wood glue.

Procedure

Cut a piece of 1/4-inch wooden dowel about 6 inches long. Make sure that both ends are flat and smooth. Obtain a flat wooden base. Position the dowel in the middle of the base. Secure it to the base with a bead of glue. Let it dry.

Slip a magnet over the dowel. Slip another one on top. If the magnets attract, remove the upper magnet. Flip it upside down so it is repelled by the magnet below. This will make the upper magnet appear to float and bounce in air. Add several more magnets. Make sure that each magnet repels the magnets on either side of it.

This is What Happens

As you've learned, like poles repel and unlike poles attract. We create a force of repulsion by positioning like poles to like poles. This force is strong enough to keep the upper magnets suspended in air.

125. Paper Flop

The aim of this easy experiment is to show the effect of speed on air pressure.

Material Required

- 2 books of equal size;
- 1 sheet of notebook paper;
- 1 drinking straw;
- Ruler.

Procedure

Position the books 4 in. (10 cm) apart on a table. Lay the sheet of paper across the space between the books. Place the end of the straw just under the edge of the paper. Blow as hard as you can through the straw.

This is What Happens

The paper flops down towards the table when air is blown under it.

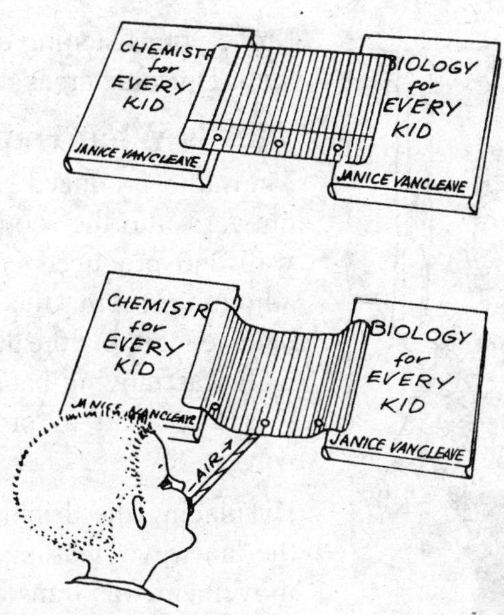

Science Behind It

Air was pushing equally on all sides of the paper before you blew through the straw. As the speed of a flow of air increases, the sideways pressure of the air decreases. Forcing a stream of fast-moving air under the paper reduces the upward pressure on the paper. The air pushing down on the paper is greater than the air pushing up, thus the paper flops down.

126. Hearing Glasses

Did you know that windows talk? Takh, tak, tak and tak. Well, not exactly. But surfaces, such as windows, vibrate to the beat of a person's voice. Several years ago, it was discovered that spies monitored the vibration of windows by using high-tech equipment. Although the movement was very slight, it could be detected by a beam of laser light. By analyzing the reflected beam, the equipment could distinguish words that were spoken in secrecy. Amazing isn't it.

Material Required

- Drinking glass;
- Wall;
- Radio.

Procedure

Find a solid wall that divides two rooms. Place a radio in one room with the volume set to a very soft level. Shut the door and go into the adjoining room.

Place a drinking glass against the wall. The rim of the glass should lie flat against the wall. While holding the glass, place your ear against the base of the glass. What do you hear?

Move your listening device to other parts of the wall. Do some areas transmit louder sounds?

This is What Happens

The radio produced sound waves that travelled outward into the room. These waves struck the wall and produced very tiny vibrations. In the adjoining room, the vibrations could not be detected without the drinking glass. Although the wall was moving back and forth, its movement was too small to produce a detectable sound wave.

By placing the drinking glass against the wall, the "air wave" was not needed. Instead, the wall's movement was transferred directly to the glass. From there, the vibrations travelled directly to the ear and were interpreted as sound.

127. Magical Drawing

Have you ever wanted to create something that blended art and science? Well, here's your chance! In this experiment, you'll get to construct a drawing that has moving parts. The parts, however, aren't magical—just magnetic.

Material Required

- Two poster boards or stiff cardboards;
- Box of crayons;
- Markers;
- Tape;
- A bunch of metal paper clips;
- Pair of scissors;
- A small magnet.

Procedure

Draw a colourful setting on a poster board. Be creative. The setting can range from a city street to a fish tank.

Identify several objects that will move around within the setting, such as helicopters, cars, or tropical fishes. Draw these objects on a separate sheet of poster board. Use a pair of scissors to carefully cut out their outlines. Colour the objects. Tape a paper clip on the underside of each cutout.

Brace the poster board upright. Position a movable cutout on the scene. Place a magnet behind the board so that it is opposite the object's paper clip. As you move the unseen magnet, the object will move within your scene. What happens to the object when you release the magnet? What happens to the object when you remove the magnet from the back of the board?

This is What Happens

Magnetic fields can easily pass through paper. When the magnet was placed behind the poster board, it attracted the paper clip. The force was so strong that the clip remained held against the upright surface.

128. Curve Ball

In a baseball game, you may have seen a pitcher throwing a curve ball. Have you ever wondered why he does that? In this experiment, we will demonstrate how a baseball pitcher throws a curve ball.

Material Required

- Golf ball, or any ball of comparable size and weight;
- String, 3 ft. (90 cm);
- Masking tape;
- Yardstick (meter stick).

Procedure

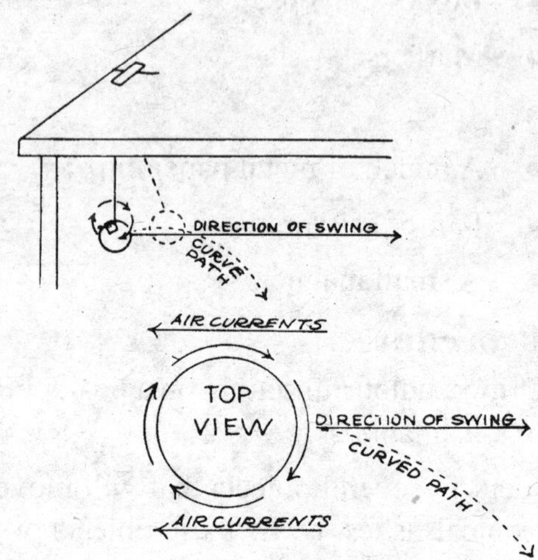

Attach the string to the ball with a strip of tape. Tape the free end of the string to the top edge of a table. The ball must be able to swing freely from the table. Lay the yardstick on the floor directly under the ball and pointing in the direction that the ball will be swung.

Twist the string about 50 times in a counterclockwise direction. Pull the ball and the twisted string back and release it. Allow the ball to swing back and forth. Use the yardstick on the floor to help you observe the direction of the swing.

This is What Happens

The ball spins in a clockwise direction as it moves forward. After a few swings, its forward motion curves to the right.

Science Behind It

The ball is moving in two directions, around on its own axis and forward. These two directions of motion cause air to flow around the ball in two different directions. As the ball spins clockwise, air is carried with the spinning ball in a clockwise direction. Air pushes in the opposite direction when the ball moves forward. Look at the diagram of the top view of the moving ball. Notice that the arrows are pointing in the same direction on one side of the ball. This indicates that the air is moving in the same direction, which causes an increase in the speed of the air on this side. The air currents are moving in opposite directions on the other side and push against each other, forcing the air to move more slowly on that side. Fast-moving air applies less sideways pressure, so the ball has less pressure on the side with the fast-moving air and more pressure on the side with the slow-moving air. The ball curves to the side with less air pressure. To throw a curve ball, a pitcher must spin the ball as he throws it forward. If he spins it clockwise it will curve to the right, but if he spins it counterclockwise the ball will curve to the left.

129. Building Big Boomer

Fireworks do it. But did you realize that you can also create a huge boom in the privacy of your own living room?

Material Required

- Corrugated cardboard (from a shipping box);
- Pair of scissors;
- Brown paper;
- Pair of scissors;
- Ruler;
- Tape.

Procedure

Cut out a triangle from brown paper. The triangle should measure 14 inches x 14 inches x 20 inches. Cut a square of corrugated cardboard. The square should measure 12 inches on a side. Use your scissors to score a straight line from one corner to the opposite corner of the square. Make sure that you only score the surface and do not cut completely through the cardboard. Place the brown paper on a flat surface. Place the cardboard over it as shown in the illustration (with scored-side up). Fold over the edge of the brown paper and secure it to the cardboard with tape. Pick up this clapper device. Fold it away from the score line, so that the paper folds inside the closed cardboard halves. Hold the free end of the clapper overhead. Fling your arm down as fast as possible. What do you hear?

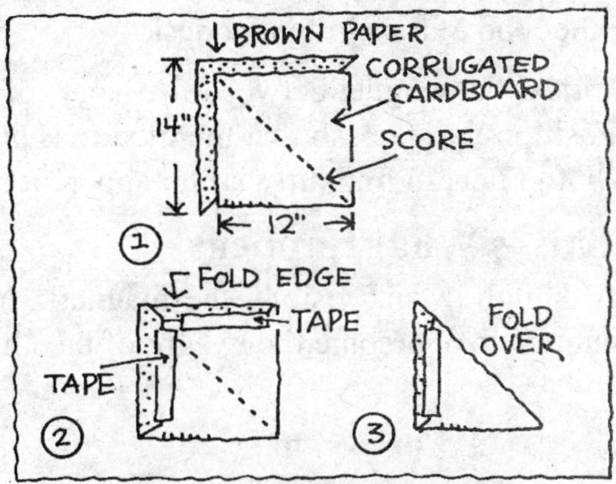

This is What Happens

As your arm dropped, the clapper opened. Air that rushed in to fill the opening space moved fast enough to create a shock wave. Although this boom was smaller than the shock waves made by supersonic jets, it was formed in a similar way.

The crack of a whip is also caused by a shock wave. The whip crack breaks the sound barrier. A snap of the wrist produces a wave that travels down the whip. As the wave moves, it increases in speed. By the time it reaches the tip, it is travelling faster than the speed of sound. This creates a shock wave that we hear as the whip crack.

130. Flying Saucers

Do you believe in Flying Saucers? Have you seen one? If so, this next experiment may be just the thing. It's an impossible sight that's made possible by a magnet's invisible force field.

Material Required

- Paper;
- Markers or crayons;
- Metal paper clip;
- 10-inch-long string;
- Tape;
- Strong and flat magnet;
- Stack of books;
- A pair of scissors.

Procedure

Draw and colour a small picture of a Flying Saucer or UFO. Cut out this object. Use tape to secure the back of this picture to a metal paper clip.

Cut a length of string about 10 inches long. Tie one end of the string to the paper clip. Tape the other end of the string to a desk.

Hide a flat magnet between the pages of a book. Place this book atop a stack of other books. Position the stack so that the picture is attracted to the magnet. When raised, the flying saucer should float in mid-air with no apparent reason for the gravity-defying behaviour.

This is What Happens

Although it isn't visible, the magnetic field exists. The hidden magnet's attraction is strong enough to overcome the weight of the clip, paper, tape, and string.

131. Making a Sprayer

You have used many types of perfumes at home, deodorants and body sprays. In this simple experiment, we will show you how a perfume sprayer works.

Material Required

- Two flexible drinking straws;
- Drinking glass;
- Scissors.

Procedure

Fill the glass with water. Cut one straw so that the top of the flexible section stands 1/2 in. (1 cm) above the water's surface. Stand this straw in the water.

Hold the second straw horizontally so that its end is pointed across the top of the other straw. Use the ridges on the straw standing in the water as a support. Blow hard through the horizontal straw.

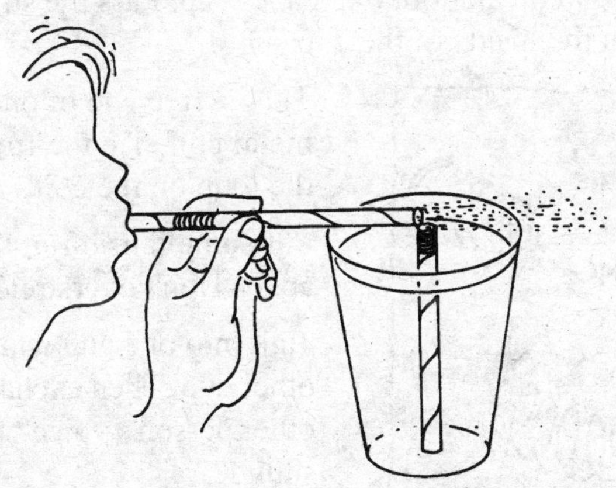

This is What Happens

Water rises in the standing straw and is blown out in a mist.

Science Behind It

The faster air moves, the lower the pressure around the flow of air. As the air from the straw moves across the top of the standing straw, the pressure inside the standing tube is lowered. Atmospheric pressure in the room pushes down on the surface of the water in the glass, forcing the water to the top of the straw, where it is blown out in a mist. Squeezing on a perfume sprayer produces the same situation. Air is forced across a tube and the perfume rises due to the reduced pressure inside and is thus sprayed outward by the moving air.

132. Alien Sounds

In space no one can hear any sort of sound! In order to transfer sound, you need matter. And, for the most part, space is empty. There is no air.

Speaking of space, this next device makes some strange, far-out sounds. It also changes the quality of your voice. Just listen.

Material Required

- Metal coil toy;
- Kite string;
- Two paper cups;
- Pair of scissors;
- Two paper clips.

Procedure

Cut two lengths of kite string, each about 6 inches long. Tie a paper clip to one end of each string. Punch a small hole in the bottom of a paper cup. Pass the string through the hole so that the paper clip remains on the inside of the cup.

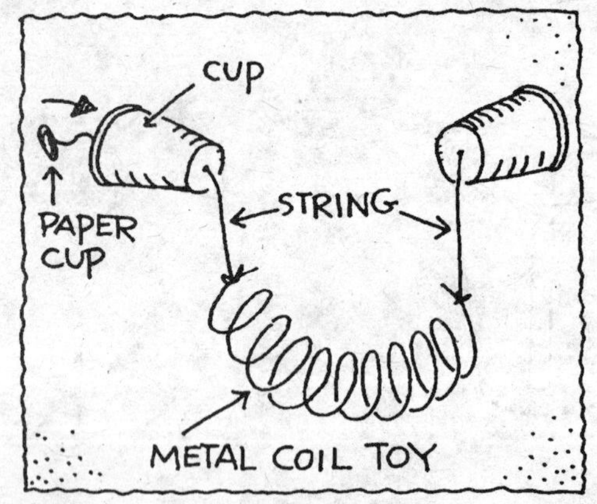

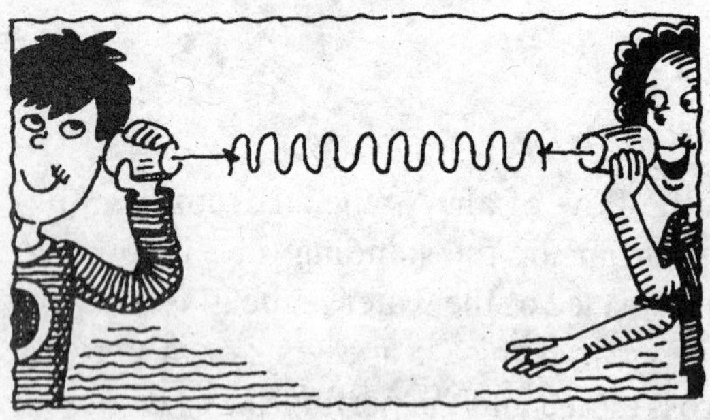

Tie the free end of one string to the end loop of a metal coil. Tie the free end of the other string to the loop on the coil's opposite end.

Put some tension in the coil by gently pulling it apart. Don't overstretch the coil!

Talk into one cup while your friend listens in the other cup. Then exchange roles. Can you hear the other person's voice? How does the coil affect the sound?

This is What Happens

When you spoke, you produced vibrations of air particles. These vibrations were collected by the cup. The movements travelled down the string and struck the coil. Not all of the vibrations emerged from the opposite end of the coil. Some of the vibrations bounced back and forth within the coil. As they rebounded, some of the sounds were transferred to the distant cup. These lagging sounds produced the weird echo-like effect.

133. Field Blockers

Now that you've had some experience with magnetic fields, it's time to make some predictions. Examine the list below and circle the materials that you believe will block a magnetic field.

- Aluminum foil;
- Wood;
- Piece of videocassette tape;
- Audio CD;
- Turntable record;
- Coins;
- Now test out your predictions by doing the following experiment.

Material Required

- Metal paper clip;
- 10-inch-long string;
- Tape;
- Wooden or plastic ruler;
- Stack of books;
- The materials listed above;
- Strong magnet.

Procedure

Cut a length of string about 10 inches long. Tie one end of the string to the paper clip. Tape the other end of the string to a desk.

Tape a magnet to the end of a ruler. Insert the ruler between the pages of a book so that the magnet is extended as far from the book as possible. Place this book on top of a stack of several books.

Position the stack so that the magnet is positioned directly above the paper clip. The clip's string should be short enough to create a gap between the clip and magnet, but long enough for the clip to remain supported by the magnet's attraction.

Place various objects directly between the magnet and the clip. Record how the magnetic field behaves.

This is What Happens

Some materials (those made of iron) will block a magnetic field. Most materials, however, do not and will allow the penetration of the field.

134. Floating Ball

How air speed affects flight? This what we are going to find out in this easy experiment.

Material Required

- Small funnel;
- Table tennis ball.

Procedure

Turn the funnel upside down. Hold the table tennis ball in the funnel with your finger. Start blowing into the narrow end of the funnel. Remove your finger from the ball as you continue to blow into the funnel.

This is What Happens

The ball floats inside the funnel.

Science Behind It

The faster the air passes by the ball, the less pressure it exerts upon the ball. The air pressure above the ball is less than the pressure under it, so the ball is held up by air. The pressure of moving air explains the upward lift on the wings of aircraft. When the air flows faster over the top of the wing than below, there is an upward push called *lift*.

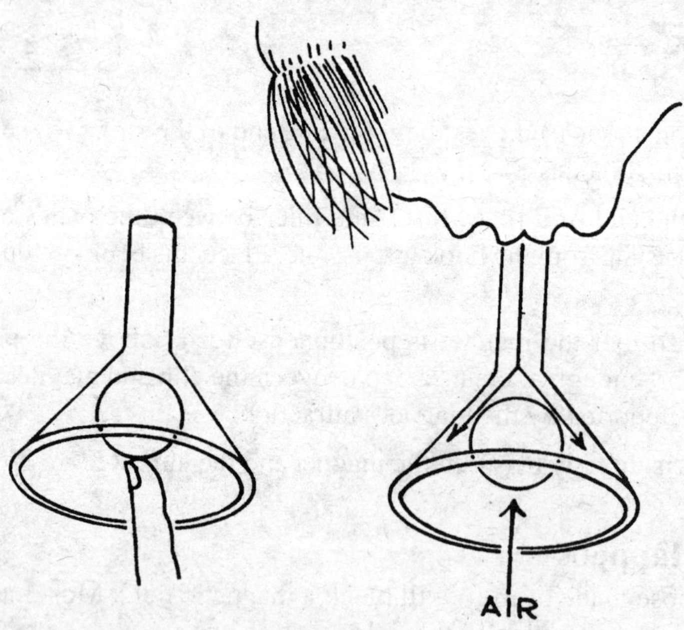

135. Magnetic Auto in Motion

Try this experiment just for fun.

Material Required

- Foam core board;
- Markers;
- A pair of chopsticks;
- Two magnets;
- Stiff paper;
- A bunch of metal paper clips;
- Tape;
- Four books of the same width.

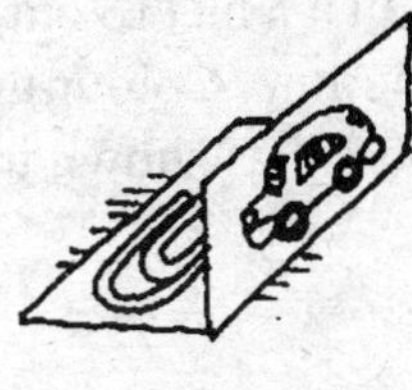

Procedure

Sketch a bird's-eye-view of a roadway onto a large piece of foam core board. The road should split, connect, and form several intersections. Include things like houses, factories, rivers, and garages along the road. Elevate the board by supporting each corner with a book.

Copy the car shown above onto a piece of stiff paper. You may wish to make other vehicles, such as fire engines, ambulances, and construction trucks. Bend the paper along the dotted line. Tape a paper clip to the upper surface of the lower fold (see illustration). Put the car on the roadway of the foam core board.

This is What Happens

Foam core does not interfere with magnetic fields. Magnets placed below the roads attracted the paper clip vehicles. As the magnets moved, they dragged the clips (along with their attached vehicles) across the foam core's surface.

136. Winding Road

You may have seen winding roads whenever you have travelled to a hill stations for holidays. The purpose of this experiment is to demonstrate that a winding mountain road is an inclined plane.

Material Required

- Sheet of paper;
- Pencil;
- Scissors;
- Ruler;
- Cellophane tape.

Procedure

Cut a 5 in. (13 cm) square from the paper. Draw a diagonal line across the square and cut across the line. Colour the longest edge of the paper triangle with the pencil. Tape the triangle to the pencil as shown in the diagram. Wind the paper onto the pencil.

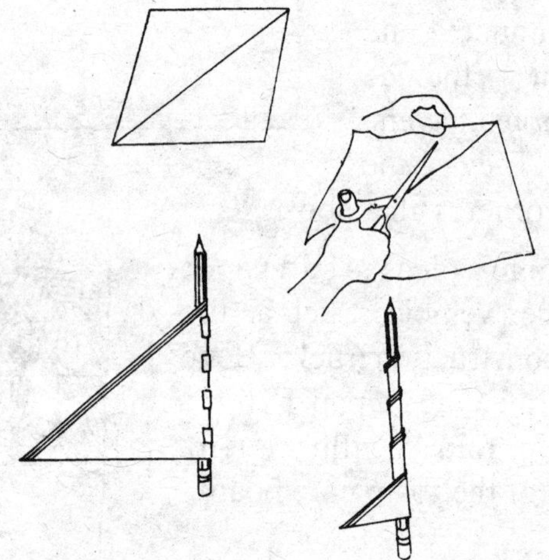

This is What Happens

The coloured side of the triangle is shaped like a ramp. Wrapping the paper around the pencil makes it look like a winding road or a screw.

Science Behind It

An inclined plane is a tilted flat surface. A winding road gradually gets higher. If the road could be unwound, it would look like the paper triangle with the road running up the long, coloured side. It is true that it would take longer to travel up the winding road than it would to walk up the mountain's side, but it takes much less effort to travel the longer distance. The inclined plane is a simple machine that makes a job easier. In this case, the job is climbing a mountain.

137. Magnetic Separator

If you were given a mixture of iron filling and salt. How can you separate these two substances?

Material Required

- Sealable plastic bag;
- Magnet;
- Plate;
- 1 teaspoon salt;
- 1/4 teaspoon iron filings.

Procedure

Add about 1/4 teaspoon of iron filings to 1 teaspoon of salt. Mix well and pour this blend onto a plate.

Seal the magnet in the plastic bag. Predict what will happen when you move the magnet through the mixture. After you make your prediction, find out what happens.

This is What Happens

The iron filings are attracted to the magnet and stick to the bag's outer surface. The salt (which isn't a magnetic material) remains behind.

138. Screw Jacks

How a screw works? This is what we are going to find out.

Material Required

- Large screw.

Procedure

Hold the head of the screw with one hand. Put two fingernails on the first ridge at the tip of the screw. Turn the head of the screw.

This is What Happens

As the screw turns, the ridges spiral downward. Your fingers stay in place and this unique winding road moves past them.

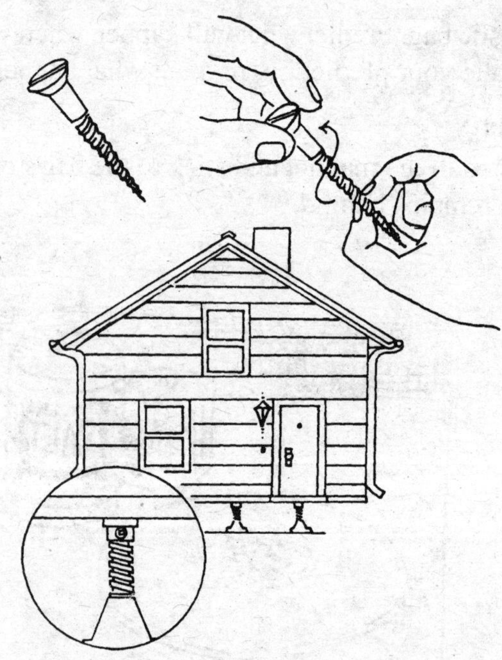

Science Behind It

The screw is an example of a simple machine called an inclined plane. Inclined means tilted, and a plane is a flat surface. The screw is like a winding ramp. The closer the treads are together, the easier it is to turn the screw. Screws are used to connect things, but they are also used to lift things; for example, screw jacks lift houses and cars. Machines are used to make a job easier.

139. Stationary Drops

Not only solid objects, like coins and pop bottles, resist changes in their movement but liquids and other fluids also have inertia? Here's an experiment, that shows how a liquid tends to stay put.

Material Required

- Mug;
- Vegetable oil;
- Water;
- Food colouring.

Procedure

Fill a mug halfway with water. Carefully pour oil into the cup so that the surface of the water becomes covered by a thin layer of oil. Place four drops of food colouring onto the oil. The drops should be positioned at the corners of an imaginary square as shown in the illustration.

While the mug remains on the table, grasp it from above. Quickly turn your wrist so that the mug spins about a quarter turn. What happens to the droplets? Can you explain your observations?

This is What Happens

Oil and water don't mix. When the oil was added to the mug, it floated upon the water's surface. The drops of food colouring entered the oil surroundings but did not flow into the water layer below. Since food colouring does not mix with oil, the dye remained as round droplets.

Although the mug spun through one quarter turn, the drops (and the mug's other liquids) remained mostly stationary. It was the inertia of these substances that kept them from moving.

140. Wedge in Work

The wedge is considered one of the five simple machines. Wedges have been used since prehistoric times to split logs and rocks; for rocks, wooden wedges, caused to swell by wetting, have been used. In this experiment, we will demonstrate how a wedge works.

Material Required

- Pencil;
- Piece of cardboard.

Procedure

Hold the eraser of the pencil on top of the cardboard. Press the pencil down against the cardboard. Press hard. Observe what happens.

Press the sharp point of the pencil against the cardboard, as hard as before.

Caution: Be sure your hand is not under the cardboard.

Observe what happens.

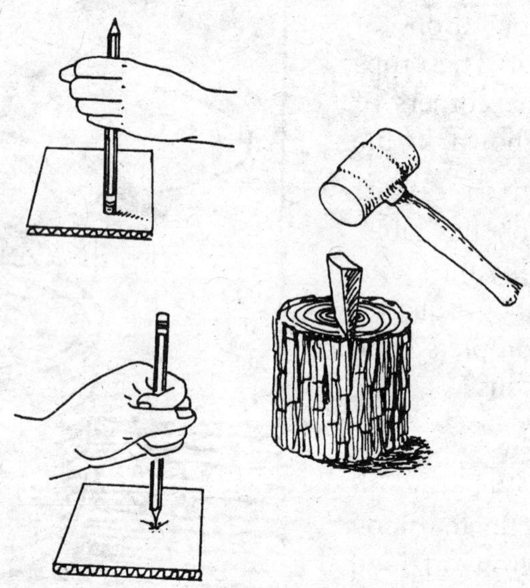

This is What Happens

The eraser presses the cardboard down, but the eraser does not stick into the cardboard. The sharp point of the pencil sticks into the cardboard.

Science Behind It

The pencil point acts like a wedge. A wedge is any material that is tapered to a thin, pie-shaped edge. The tapered, pointed end of the pencil acts as a wedge and sticks into the cardboard. The point of the pencil enters the paper first and makes a path for the larger part of the pencil to follow. An example of a simple machine used to split logs is a wedge.

141. Spinning Eggs

If you were given two eggs, one is boiled another is not, now the problem is that how can you find out, which is fresh or hard-boiled?"

Material Required

- Hard-boiled egg;
- Fresh egg.

Procedure

Place a fresh and hard-boiled egg side by side. Then challenge a friend to identify the fresh and hard-boiled eggs by observing their spins.

While on its side, give one egg a whirl. Gently stop it in the middle of its spin. Release the egg. What happens?

Repeat this action with the other egg. How do these eggs differ in their behaviour?

This is What Happens

Inertia. It's that word again! To understand how inertia puts the spin on things, we must first examine the contents of these eggs.

The fresh egg has a liquid interior. While it spins, both its outer shell and liquid interior moves. When the egg is stopped, only the exterior comes to a halt. The liquid on the inside keeps spinning. So if the stopped egg is released, the spinning yoke gets the egg turning again.

The hard-boiled egg is solid on the inside. When it spins, both shell and hard insides turn. When it stops, everything stops completely. If the stopped egg is released, it will not start to spin again.

142. Inclined Plane

The purpose of this experiment is to determine how an inclined plane makes work easier.

Material Required

- String, 12 in. (30 cm);
- Rubber band;
- 2 books;
- Glue bottle, 8 oz. (236 ml);
- Ruler.

Procedure

Tie the centre of the string around the top of the glue bottle. Attach the rubber band to the string. Raise the end of one book and rest it on top of the second book. Place the glue bottle on the table near the flat book.

Hold the rubber band and lift the bottle of glue straight up and place it on the book. Measure with the ruler how much the rubber band stretches. Lay the glue bottle at the bottom of the inclined book.

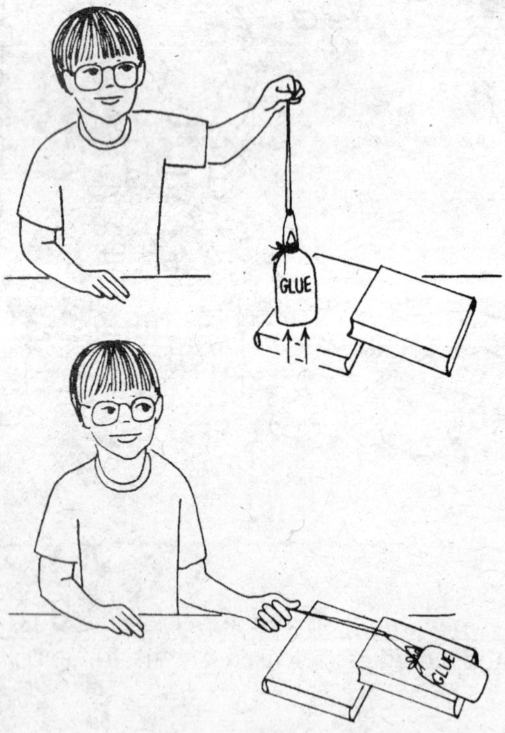

Hold the rubber band and pull the bottle until it reaches the height of the flat book. Measure with the ruler how much the rubber band stretches.

This is What Happens

The rubber band stretches the most when the bottle is lifted straight up.

Science Behind It

You lifted the glue bottle to the same height each time, but lifting it straight up took more effort because you were holding the entire weight of the bottle. The books form a simple machine called an inclined plane that helps to support some of the weight of the bottle. You had to pull the bottle up the incline a longer distance to reach the height of the book, but it took much less effort, as shown by the rubber band that stretched less.

143. Midair Catch

Magicians, jugglers, and other performers often have a good understanding of physical science. Some understand acceleration and use it to juggle clubs. You can do the same with some coins too.

Material Required

- A bunch of coins.

Procedure

Balance a stack of coins on your elbow. Snap your arm forward and catch the coins in midair!

This is What Happens

Like many tricks, this one is not as difficult as it appears. That is because you have a helper: inertia. As you snap your arm forward, the coins are left unsupported in midair. Since they have inertia, they do not move sideways or drop too quickly. Instead, their drop begins slowly. By the time your wrist comes around, the coins have not had enough time to gain much speed. Instead, their relatively slow movement makes them an ideal target for a midair snatch!

144. Wheel Pump

The aim of this experiment is to show how a simple wheel can be used as a pump.

Material Required

- Eye dropper;
- Glass or wide-mouthed jar with a 4 in. (10 cm) opening;
- Poster board;
- Round toothpick;
- Scissors;
- Helper.

Procedure

Cut a circle with a 4 in. (10 cm) diameter from the poster board, using the glass to outline the circle. Insert about one-third of the toothpick through the centre of the circle to form a top.

Fill the eye dropper with water. Ask your helper to spin the paper top. Hold the eye dropper above the top and drop water onto the spinning disk. Observe the movement of the water.

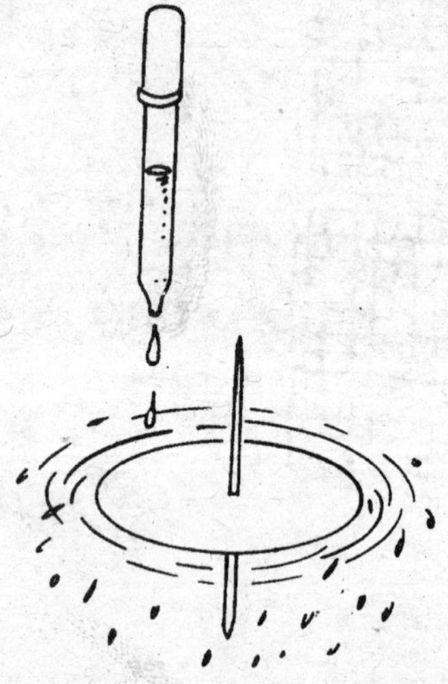

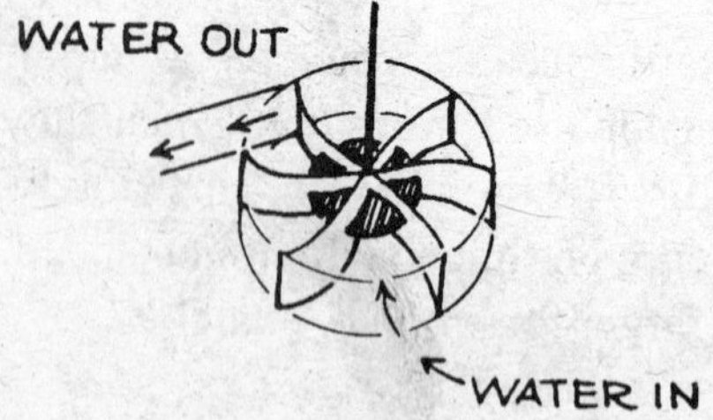

This is What Happens

The water sprays out in all directions.

Science Behind It

The falling water is thrown away from the spinning disk. If you could slow down the motion of the water, you would see that the water leaves the spinning top in a straight line. There is an outward pull on all spinning materials; if free to move, they are thrown off in a straight line. This knowledge was used to design a simple machine to pump water. Water flowing onto a spinning metal disk creates a wheel pump that very effectively pumps water from one spot to another.

145. Juggling Marbles

Now that you've learned how to catch a stack of coins, how about catching marbles in a mug? As you may discover, this trick can get frustrating, especially, if you don't uncover the secret of the technique.

Material Required

- Several marbles;
- Mug.

Procedure

Hold the mug's handle with your pinkie, ring, and middle finger. This leaves your index finger and thumb free. Using only these two fingers, pick up a marble. With an upward arm motion, toss the marble into the air (but keep holding the mug). Quickly position the mug beneath the falling marble. Great catch!

Now, it gets tougher. While grasping the mug with the marble still inside, pick up a second marble with your index finger and thumb. Now try throwing this second marble, so that it joins the first in the mug.

Magic Words

If you were lucky enough to get the second marble in (using this same technique), try tossing and catching a third one.

This is What Happens

The first marble is the easy one to catch. As it rises, you position the mug directly beneath it. Gravity does the rest.

The next marble is more difficult to snare (while keeping the first one within the mug). If you try the same technique, the captured marble keeps jumping out!

The best way to capture the second marble is using a "drop-n-slide" technique. Don't toss the second marble upward. Instead, release your grip on this marble. As it falls, quickly position the mug in its downward path. As you'll discover, the mug only needs to drop about an inch for the marble to clear the rim.

146. Levers

All early people used the lever in some form, for moving heavy stones or as digging sticks for land cultivation. Balance beams for weighing were probably used in Egypt c. 5000 BC; they consisted of a bar pivoted at its centre with weights on one end balancing the object on the other. As early as 1500 BC people were raising water and lifting soldiers over battlements using the swape or shadoof, a long lever pivoted near one end with a platform or container hanging from the short arm and counterweights attached to the long arm. This experiment demonstrates the effectiveness of a lever.

Material Required

- 4 books;
- 2 pencils.

Procedure

Stack the books. Put your little finger under the edge of the bottom book in the stack and try to lift the books.

Place one pencil under the edge of the bottom book in the stack. Place the second pencil under the first pencil near the book. Push down on the end of the second pencil and try to lift the books.

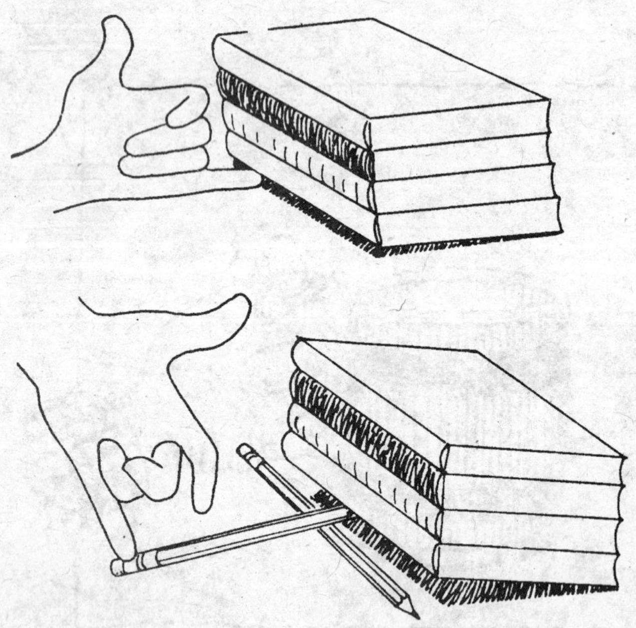

This is What Happens

It is very difficult to lift the books with your finger alone, but easy when two pencils are used.

Science Behind It

The pencils form a lever. One of the pencils acts as a fulcrum (a point of rotation), and the second pencil is the lever arm. As the distance from where you push down to the fulcrum increases, the easier it is to lift the object on the opposite end. Levers are simple machines that multiply the force that you apply. This makes moving or lifting large objects easier.

147. Shooting Coins

It's multiple guess time! Imagine a stack of four coins set on a flat tabletop. Suppose you were to shoot a fifth coin at the stack so that it hits the bottom coin dead centre. What do you think will happen to the stack?

The fifth coin will bounce off the stack

The fifth coin will replace the bottom coin

The fifth coin will stop in front of the stack, but the bottom coin will shoot out.

Material Required

- Five coin (same size).

Procedure

Stack four coins on a slick, flat surface. Place a fifth coin several inches away. Flick the coin at the stack so that it strikes the stack squarely in the middle of the bottom coin. The flip doesn't have to be very powerful, just dead on centre.

This is What Happens

Off-centre Shots: The stack crumbles as this adventure goes down the drain. Glancing blows don't work.

Dead Centre Shots: The energy of the collision is transferred to the coin at the bottom of the stack. As it moves off, the three coins above it fall straight down because of their inertia. The flicked coin transferred all of its moving (or kinetic) energy upon the collision to the bottom coin but it lacked the "extra" power needed to take its place beneath the remaining stacked coins. Therefore, the bottom coin flew off, but the fifth coin remained in front of the now three-coin stack.

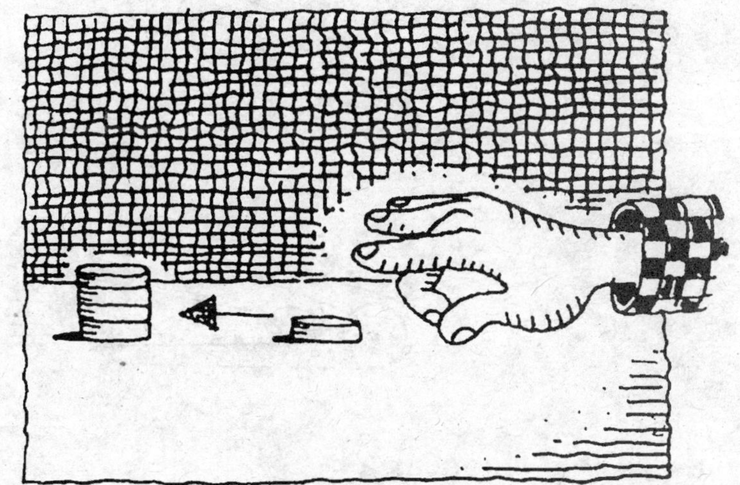

148. Second Class Lever in Help

The aim of this simple experiment is to demonstrate a second-class lever.

Material Required

- 2 round toothpicks.

Procedure

Place a toothpick across the back of your middle finger at the first knuckle and under the first and third finger. Try to break the toothpick by pressing down with your first and third fingers. Move the toothpick closer to the tips of your fingers. Again push down with your fingers to try to break the toothpick.

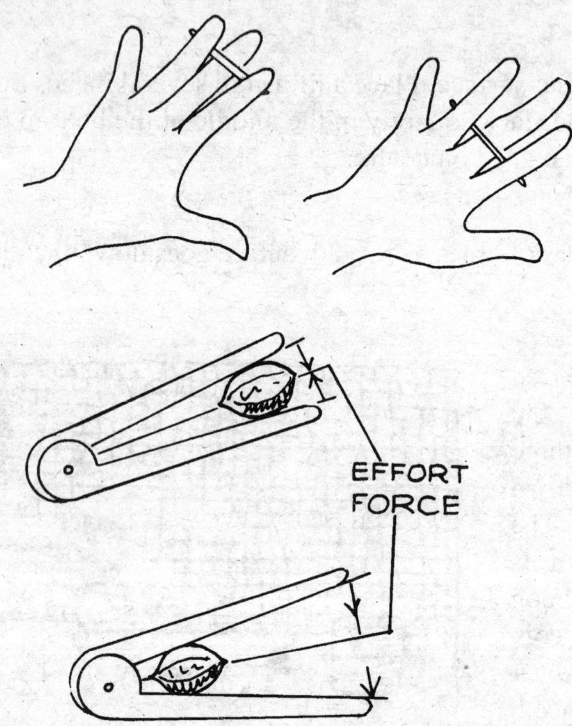

This is What Happens

It is very difficult or impossible to break the toothpick when it is at the ends of your fingers.

Science Behind It

Your fingers act as a second-class lever similar to a nut cracker. The point of rotation or fulcrum is where the fingers join the hand. When the toothpick is placed furthest from the fulcrum, the force needed to break the toothpick is greatest. Placing the toothpick close to the fulcrum requires less effort—force—to break the wood.

149. Watery Trick

Have you ever seen a science fiction movie that takes place in a spinning space station? If so, chances are that the astronauts weren't weightless. Instead, the rotation produced some sort of artificial gravity that kept the astronauts against the station's edge.

In this experiment, we will observe a down-to-earth spinning effect.

Material Required

- Small plastic sand bucket;
- 2-foot long kite string;
- A cup of water.

Procedure

Find a sand bucket. Cut a section of kite string about 2 feet long. Tie the string to the arch of the handle. Add about a cup of water to the bucket. Test the knot to make sure that it won't slip out.

Go some place outdoors or to an area where spilling water won't be a problem. Make sure you have plenty room around you so that you won't hit anyone or anything.

Now grasp the free end of the string and lift the bucket. Start swinging the bucket in a circle using a steady and gentle movement. Increase the force and angle of the swing so that the container "climbs" higher and higher. Keep going until the bucket loops completely upside down. What happens to the water?

This is What Happens

The spinning motion produced a "force-like" effect that kept the water within the bucket. Since the effect of the spin was stronger than the attraction of gravity, even when the bucket spun through an upside down position, the water didn't spill out.

150. Fine Position

The purpose of this experiments, is to determine if the position of the lever's fulcrum affects the lever.

Material Required

- Ruler;
- Pencil;
- 30 pennies.

Procedure

Make a lever by laying the ruler across the pencil. Move the pencil under the 4 in. (10 cm) mark on the ruler. Place 10 pennies between the end of the ruler and the 1 in. (2.5 cm) mark. Add and record the number of pennies needed on the opposite end of the ruler to lift the 10 pennies. Move the pencil to the 8 in. (20 cm) mark. Again position 10 pennies on the end of the ruler.

Add and record the number of pennies needed on the opposite end of the ruler to lift the 10 pennies.

This is What Happens

It took many more pennies to lift the 10 pennies when the pencil was at the 8 in. (20 cm) mark.

Science Behind It

The ruler and pencil form a simple machine called a lever. The pencil acts as the fulcrum that the lever rotates around. The effort is the amount of force needed to make the object move. In this experiment, the effort was the stack of pennies needed to lift the 10 pennies. The longer the distance from the point of effort to the fulcrum, the less effort needed to move the object on the other end. You saw this on a See-saw when the heavy person moves experience the centre in order to balance the lighter person sitting on he other end. The lever can make lifting or moving an object an easy task if the fulcrum is placed close to the object to be moved.

151. Snapping String Trick

Here's another great trick. Tell your audience that you can snap the thread either above or below the fishing sinker. Let them decide on where they want you to break the thread. Then snap the string according to their wishes.

Material Required

- 2 ounce fishing sinker;
- Fine sewing thread.

Procedure

Break off two pieces of thread, each about 6 inches long. Tie both threads to the "eye" of a fishing sinker. Tie the free end of one thread to a secure support so that the sinker hangs freely.

Grab hold of the other free end of the hanging thread. Gradually increase the force of your pull. What happens? Where does the thread break?

Replace the broken thread. Instead of a gradual pull, this time give the lower thread a quick snap. Where does the thread break? Can you explain your observations?

This is What Happens

When you pulled down slowly, you overcame the inertia of the sinker. In fact, the sinker's weight added to your pulling force to place a greater tension in the thread above the sinker than in the thread below the sinker. This caused the thread to snap above the weight.

When you snapped the thread quickly, the inertia of the sinker worked against the pull. Its resistance opposed the pull. This created a greater stress beneath the sinker, causing the break to occur in this lower segment.

152. Wheel and Axle

Axle combined with the wheel, in its earliest form it was probably used for raising weights or water buckets from wells. In this experiment demonstrates how machines make work easier.

Material Required

- Empty thread spool;
- 2 pencils;
- String;
- Scissors;
- 2 paper cups;
- 20 pennies;
- Marking pen.

Procedure

Stick the pointed end of one pencil into each end of the empty thread spool. Be sure the pencils fit snugly and do not slide. Suspend the pencils and spool from a table edge with two loops of string. Tape the loops to the table, making sure the pencils are level. Use the tape and pen to label the cups A and B.

Punch two holes at the top of each cup and attach one 24 in. (60 cm) string to each paper cup as shown in the diagram. Tape cup A's string to a pencil. Turn the pencils away from you to wind all of the thread onto the pencil. Tape cup B's thread to the outside of the thread spool. Turn the pencils towards you to wind cup B's thread onto the spool. Place 10 pennies in cup A. Cup B should be at its top position. Add pennies to cup B one at a time until it starts to slowly move downward. Observe the distance that both cups move.

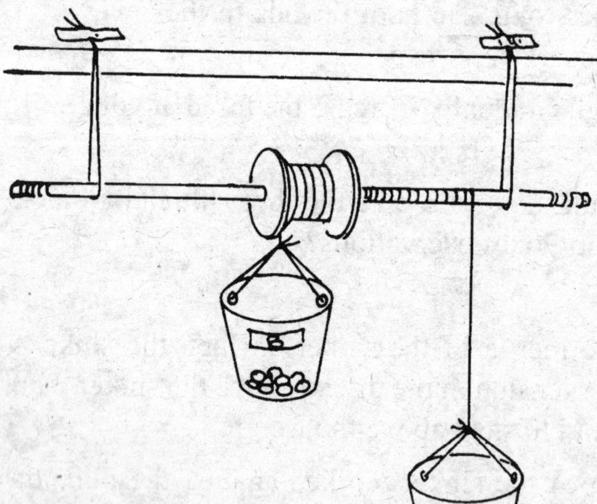

This is What Happens

It takes 6 to 7 pennies in cup B to raise the 10 pennies in cup A. Cup B moves a greater distance than cup A.

Science Behind It

The distance around the spool is farther than the distance around the pencils. This makes the cup attached to the spool move farther every time the spool and the pencils revolve. Machines make work easier. This spool and pencil machine, called a wheel and axle, needs the force of only 6 to 7 pennies to lift the load of 10 pennies.

153. Rotating Rocket

Every object has a special point called its centre of gravity. If you support an object so that it is allowed to spin freely, it will eventually come to rest so that its centre of gravity is directly below the point of support. If that's a bit complicated for you, this experiment will hopefully set things straight for you.

Material Required

- Heavy stock paper or cardboard;
- Marker;
- 6-inch-long kite string;
- Heavy stock paper;
- Pair of scissors;
- Washer;
- Pushpin.

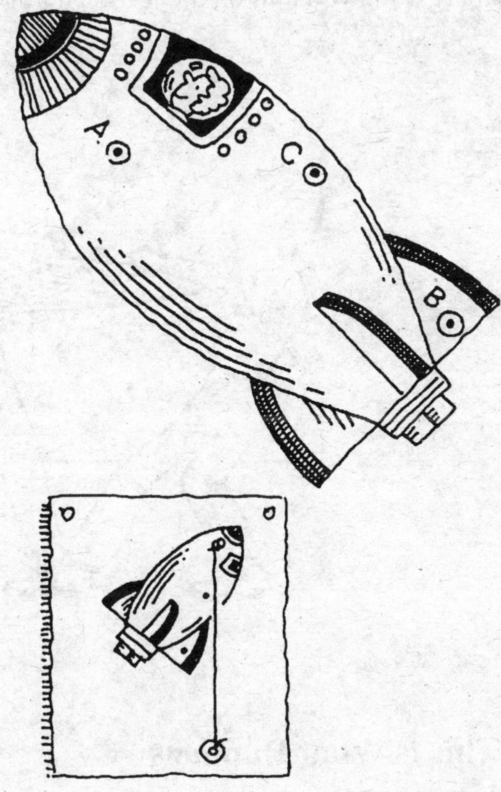

Procedure

Trace or copy the picture onto a sheet of heavy stock paper or cardboard. Cut out the drawing.

Tie a washer to a 6-inch long kite string. Press a pushpin through the circle marked A into a wooden support. The pin shouldn't be snug. It should be loose enough to allow the rocket to freely rotate.

Tie the free end of the string to the pushpin shaft. Use a marker to trace the path that the string makes. Repeat this step by tracing the lines made when the pin is placed into points B and C. Do the three lines intersect? What's so special about this point?

This is What Happens

Everything tends towards stability. The stability (balance) of the rocket was affected by the irregular distribution of weight. When the pin was placed into point A, the rocket was free to rotate into its most stable position. The drawing came to rest so that its centre of gravity was directly below the pin (point of support).

When the pin was placed at the two other points, the same thing happened. The rocket came to rest so that its centre of balance was directly beneath the pin. The three lines intersected at the rocket's centre of gravity.

154. Tug of War

The purpose of this experiment is to demonstrate how easily things are moved with a machine.

Material Required

- Two brooms;
- Rope or strong cord, 9 ft. (3 m);
- 2 helpers.

Procedure

Tie the rope to one broom handle. Wrap the rope around the broom handles three times while they are being held about 20 in. (50 cm) apart. Have your helpers try to keep the broom handles apart while you pull on the loose end of the rope.

This is What Happens

You can move the broom handles together even though your helpers are trying to keep them apart.

Science Behind It

The brooms and rope act as a pulley system. Your force is multiplied by the number of ropes attached to the brooms; therefore, you have about five times the effort or force that is being exerted by each of your helpers.

155. Floating Flavours

How can you tell a regular soft drink from its diet form? Most likely, you can taste the difference. Although manufacturers of artificial sweeteners like to believe that their "sweetening" chemicals taste just like the real thing, they don't. Most people can detect a chemical aftertaste. There's another difference too. This one, however, is seen, not tasted.

Material Required

- A can of regular (sweetened with sugar) soft drink;
- A can of diet soft drink;
- Fish tank or large bowl filled with water.

Procedure

Place a can of regular soft drink in a fish tank or large bowl filled with water. Does it float or sink? Now place a can of diet soft drink in the bowl. What happens to this can?

This is What Happens

Floating is all about the balance of forces. The downward force that causes something to sink is called weight. The weight depends upon the amount of matter "stuffed" in an object. The upward force that causes something to rise is called buoyant force. The buoyant force depends upon the amount of space the object takes up.

When the weight is greater than the buoyant force, the object sinks. When the buoyant force is greater than the weight, the object floats. When the weight and buoyant force are equal, the object remains at the same level (neutral buoyancy).

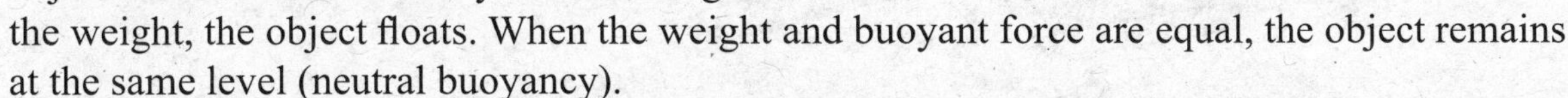

Regular soft drink is a solution of carbonated water, flavouring, and plenty of sugar. Since the soft drink's sugar is dissolved, you can't see it. You can, however, certainly taste it.

Like regular soft drink, the diet form contains carbonated water and flavouring. The sugar, however, is replaced by a small amount of a artificial sweetener. The taste of the substitute is so strong that very little is needed.

Cans of regular and diet soft drink are almost neutral in buoyancy. The extra sugar in the regular soft drink, however, makes this can, lightly heavier. The diet soft drink, although it has a small amount of artificial sweetener, lacks the sugar and is slightly lighter. Although the difference is small, it is significant enough to cause the can filled with the regular soft drink to sink, while the can filled with the diet soft drink floats.

156. Crash Landing

The purpose of all this effort is to demonstrate that moving objects have inertia.

Material Required

- Piece of modeling clay, size of a walnut;
- 2 rulers;
- Small toy car that can roll on the ruler;
- Masking tape;
- Pencil;
- 2 books about 1 in. (2.5 cm) thick.

Procedure

Raise one end of the ruler and place it on one of the books. Tape the other end of the ruler to a table. Tape the pencil perpendicular to and about two car lengths from the end of the ruler. Make a clay figure similar to a snowman.

Flatten the bottom of the clay figure and gently sit it on the hood of the toy car. You want the clay figure to fall off the car easily, so do not press the clay against the car.

Position the car with its clay figure at the top of the raised ruler. Release the car and allow it to roll down the ruler and collide into the pencil. Use the second ruler to measure how far the clay figure falls from the car. Repeat the procedure several times before raising the ruler by adding the second book. Keep a record of how far the clay figure falls from the car.

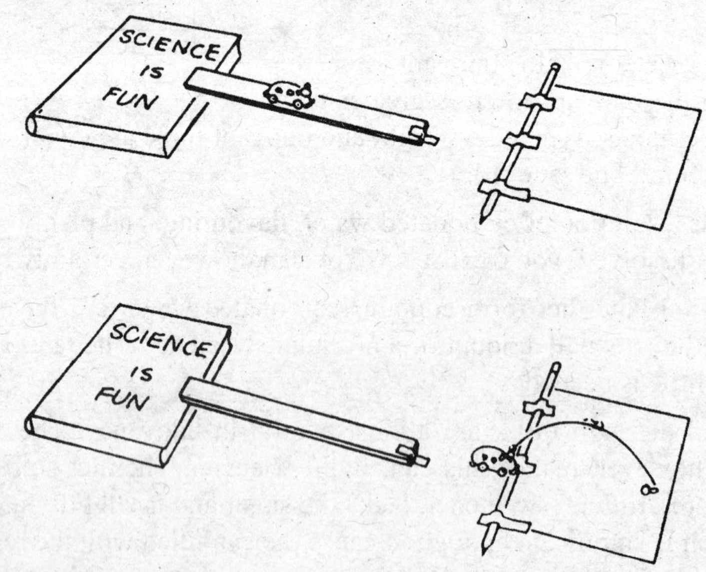

This is What Happens

The car with the clay figure moves down the ruler. The car stops when it hits the ruler, but the clay figure continues to move forward. Raising the ruler causes the clay figure to fall farther from the car.

Science Behind It

As the car rolls down the ruler, its speed increases. The clay figure has the same speed as the car. When the car hits the pencil, the force of impact stops the car, but the clay figure is free to continue moving forward until some force stops it. Raising the height of the ruler causes the car to reach a higher speed before it hits the pencil, and so the clay figure also moves at a higher speed. The faster the clay figure moves, the farther it flies before the forces of gravity and the air molecules bring it to a stop.

The car and clay figure both have inertia, a resistance to a change in motion. Once started, both continue to move until some outside force acts against them, causing them to stop. The pencil stopped the car's motion, and the air molecules slowed the clay figure's forward motion as gravity pulled the clay figure down.

157. Dead Sea Effect

The water that fills Dead Sea contains a great deal of dissolved salt. This salt makes the water "extra" buoyant. It is so buoyant that it's nearly impossible for a person to sink! But you don't have to go to Dead Sea in Israel to observe extra buoyancy. You can produce this effect right in your kitchen. Here's how.

Material Required

- Raw egg;
- Spoon;
- 1/2 cup salt;
- Tap water;
- Two tall plastic containers;
- Measuring cup.

Procedure

Fill two tall plastic containers 3/4 full with water. Add about 1/2 cup of salt to one of the containers. Mix thoroughly until the salt dissolves and then add more salt. Keep adding salt until no more salt can be dissolved. Don't worry about adding too much!

Carefully place a raw egg into the container filled with pure tap water. What happens to the egg? Now remove the egg and place it into the container filled with salt water. What happens to the egg this time?

This is What Happens

As you've learned, floating is the delicate balance between weight and buoyant force (B.F.). In this activity, you didn't change anything about the object. Instead, you changed the surrounding environment.

The strength of a B.F. is dependent upon how much "stuff" is dissolved in the solution. The more dissolved materials, the greater the B.F. When you added salt to the tap water, you not only made the solution saltier, but you increased the amount of B.F. it could produce.

When the raw egg was placed in tap water, it sunk. The egg's weight was slightly greater than the B.F. it generated. When this same egg was placed in salt water, it floated. Although the egg's weight stayed the same, the buoyant force was greater. This greater B.F. offset the egg's weight and prevented it from sinking.

158. Weight and Inertia Relation

Does weight has any thing to do with inertia. This is what we are going to find out in this easy experiment.

Material Required

- Plastic soft-drink bottle, 2 qt. (2 litre);
- Rubber band;
- String, 12 in. (30 cm);
- Scissors;
- Ruler.

Procedure

Tie the string to the rubber band. Place the rubber band around the bottom of the bottle. Pull on the string until the bottle starts to move. Measure the amount that the rubber band stretches.

Fill the bottle with water. Pull on the string until the bottle moves and again measure the amount that the rubber band stretches.

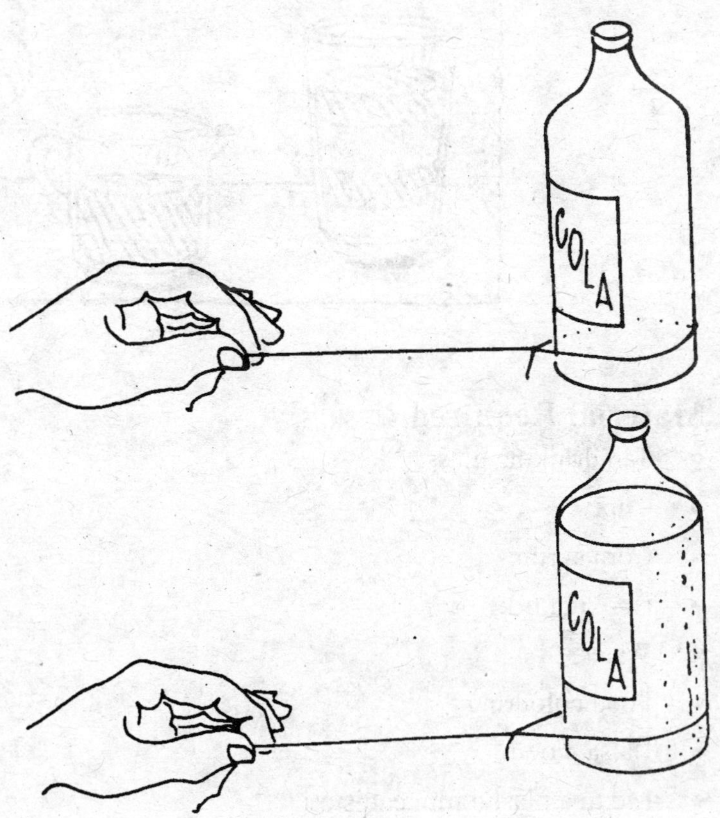

This is What Happens

The rubber band stretches very little when moving the empty bottle, but stretches much more when the bottle is filled with water.

Science Behind It

Inertia is the resistance to motion. The rubber band stretched very little when moving the empty bottle because the bottle has very little resistance to being moved. The water-filled bottle heavier and resists movement more than does the empty bottle. As a result, the rubber band stretched more when moving the heavier, water-filled bottle. As weight increases, *inertia* increases.

159. Destiny in Density

So far, we've talked about floating without mentioning the density. We've explained floating in terms of a balance between buoyant force and weight. There is, however, another way of presenting buoyancy. It has to do with a concept called density.

Density refers to the concentration of matter. Solutions that have a great deal of dissolved materials are denser than those that have nothing dissolved in them. For example, salt water is denser than plain old tap water. Likewise, a one-quart solution with 1/2 cup of dissolved salt is denser than a similar solution with only 1/4 cup of dissolved salt.

Now that you understand density, here's how it fits in with floating. When two substances are mixed together, the denser material sinks. Chocolate syrup sinks in milk. Balsa wood floats on water. Carbon dioxide shot from a fire extinguisher sinks in air.

Material Required

- Tall drinking glass;
- Cup;
- Corn syrup;
- Cooking oil;
- Water;
- Food colouring;
- Balsa wood;
- Lid to a plastic film canister;
- Piece of hard sugar candy;
- Coin.

Procedure

Carefully pour several ounces of corn syrup into a tall drinking glass so that the corn syrup forms a layer several fingers thick. Pour about 8 ounces of water into a small, clean cup. Next, add several drops of food colouring to the water (it makes the layers look more dramatic). Tilting the glass, carefully pour this coloured water into the drinking glass so that the water flows down the inner side of the container (and doesn't splatter). Save a small amount of the coloured water for later. After the water has been added, introduce the cooking oil in the same manner. Stop when the oil forms a layer that is several fingers thick.

Observe how the layers interact. Do they mix? Do they stay perfectly separated? Now add a piece of balsa wood, the lid to a plastic film canister, a piece of hard sugar candy, and a coin to the mixture. Which level do these materials float?

This is What Happens

The mixture that you just formed has three layers. The layers remain separated due to their difference in density. Eventually some mixing (mostly between the corn syrup and dyed water) will occur. Until then, this remains a three layered mixture. Since the mixing doesn't occur in an atomic level, this mess is NOT a solution. The layers stay separated because each has its own density. The corn syrup has the greatest density and remains at the bottom. The coloured water has a medium density and remains in the middle. The oil least dense and remains afloat.

Balsa wood has a density less than oil. Therefore, the wood scraps floated on top of the oil. The plastic from a film canister is less dense than water, but more dense than oil. You can tell that by observing the layer at which it floats. Hard sugar candy is denser than water, but less dense than corn syrup. The coin is the densest one of them all. It sinks to the bottom of the container.

By now, you may be saying to yourself, "Hey, wait a second. This experiment is rigged! The corn syrup stayed on the bottom, because it was poured in that layer. Likewise for the water and oil."

You are partly right. Yes, the liquids were poured in order to help produce distinct layers. Their position, however, was determined by their density. To prove this, repeat the mix but this time start with the oil first.

160. Stopped But Still Moving

The purpose of this experiment is to demonstrate that an object continues to move due to inertia.

Material Required

- 5 books;
- Chair or cart with rollers.

Procedure

Stack the books on the edge of the chair's seat. Push the chair forward then quickly stop the chair.

This is What Happens

The books move forward and fall to the floor.

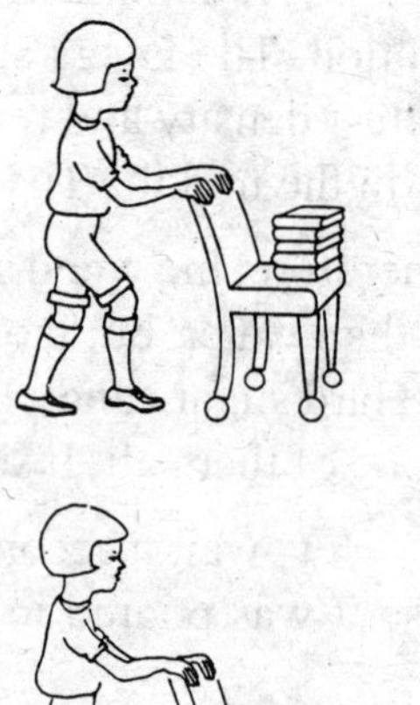

Science Behind It

Inertia is a resistance to any change in motion. A moving object remains in motion until some force stops it.

The books are moving at the same speed as the chair. They are not attached to the chair. Therefore, when the chair stops, the books continue to move forward. They would continue to move forward in the air until hitting some other object except that the force of gravity pulls them down. Air molecules also are hitting against the books, slowing their forward motion.

161. Forming Undercurrents

Powerful currents exist beneath the ocean's surface. Unlike the surface variety, these unseen currents are not created by the wind. Instead, their movement is driven by differences in density. In fact, these subsurface currents are called density currents.

When salt water evapourates or becomes trapped as ice, its load of dissolved salts remains behind. This "leftover" salt dissolves in the remaining water to make it denser. If unmixed, this dense water sinks through the less saline water below. As it sinks, the water spreads out, forming the sideways movement of a density current.

Material Required

- Salt;
- Blue and red food colouring;
- Ice;
- Spoon;
- Two cups;
- Two clear plastic pitchers.

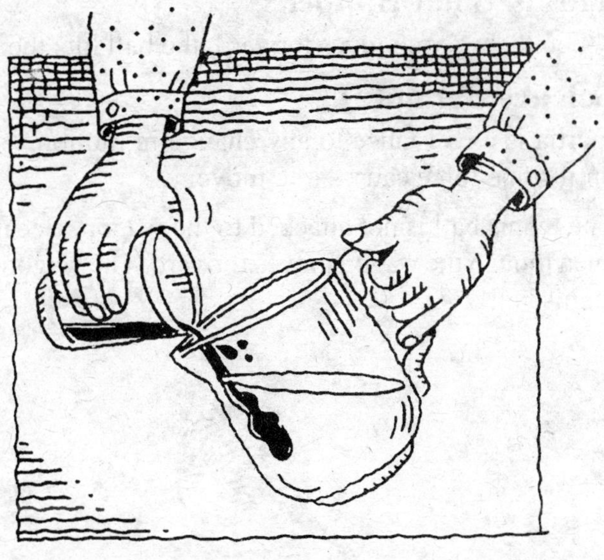

Procedure

Fill a cup halfway with ice water. Add several drops of blue food colouring.

Fill a second cup halfway with tap water. Dissolve as much salt as possible into it. Add several drops of red food colouring to this saline water.

Fill a large pitcher halfway with room temperature water. Slowly pour the dyed ice water into this pitcher. Tilt the pitcher so that the ice water runs down its inner side. Observe what happens when the ice water meets the room temperature water. Does the ice water float or sink? Is ice water more dense or less dense than room temperature water?

Fill a second pitcher halfway with room temperature water. Slowly pour the dyed salt water into this pitcher. Tilt the pitcher so that the salt water runs down the inner side of the pitcher. Observe what happens when the salt water meets the tap water. Does salt water float or sink? Is salt water more dense or less dense than tap water?

This is What Happens

When water is cooled, its tiny particles (molecules) slow down. Their slower speed allows the molecules to pack closer together. The result is an increase in density. Therefore, when cold water is poured into room temperature water, it sinks to the bottom of the container.

162. Inertia Again

In this experiment, we will exhibit that due to inertia an object remains stationary.

Material Required

- Wagon or cart;
- Tennis ball.

Procedure

Place the tennis ball in the centre of the wagon's bed. Quickly move the wagon forward.

This is What Happens

When the wagon moves forward, the ball hits the back of the wagon.

Science Behind It

Inertia is a resistance to any change in motion. An object that is stationary remains that way until some force causes it to move.

The tennis ball is not attached to the wagon. Because of the ball's inertia, it remains stationary even though the wagon moves forward. The wagon actually moves out from under the stationary tennis ball.

163. Floating Stones

In Ramayana TV serial, you have seen stone with Rama written on them floated in sea water. Where would you expect to find a rock that floats? In a fantasy story? In a science fiction movie? Try a fancy store. Strange, but true: Floating rocks are sold at most fancy stores!

Material Required

- Pumice (available at fancy or cosmetic stores);
- Small container filled with water.

Procedure

Take a piece of pumice and place it in a small container filled with water. Does the rock sink or float?

This is What Happens

Pumice is a volcanic material that looks and feels like hardened foam. It is formed when molten rock, which was filled with boiling gas, cools very quickly. As it becomes solid, the rock material traps gas bubbles.

The air pockets within the pumice keep it light. In fact, there is so much trapped air that if you averaged the density of the solid rock material and the air spaces, you'd have an overall density that is less than water. Hence, pumice floats!

164. Smack

How forces affect inertia? In this experiment, we will seek the answer to this question.

Material Required

- Drinking glass;
- Index card;
- Clothespin.

Procedure

Place the index card over the mouth of the glass. Place the clothespin on top of the card so that it is centred over the glass. Quickly and forcefully thump the card straight forward with your finger. Repeat the experiment several times.

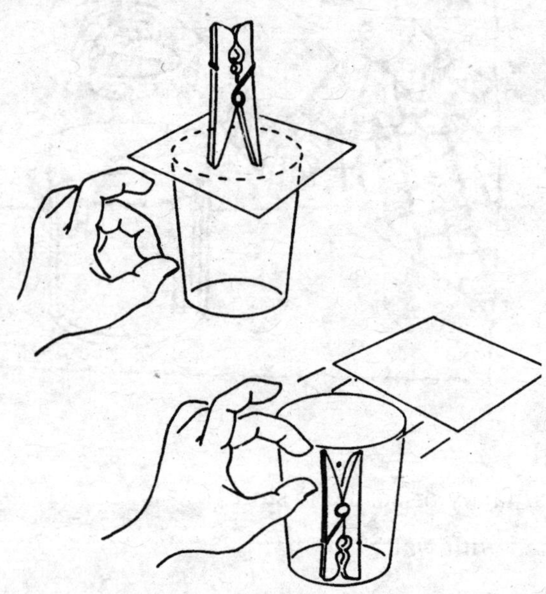

This is What Happens

The clothespin falls straight into the glass about half of the time; the other half of the time it flips over, landing upside down in the glass.

Science Behind It

Your finger applies force to the card, moving it forward. The card moves so quickly that it translates very little force to the clothespin. The pin falls straight down due to the pull of gravity when the card no longer supports it. If you do not hit the card straight forward with enough force, it pulls the bottom of the pin forward and gravity pulls the top of the pin down, causing the pin to flip before it lands.

165. Faking Pumice

Other materials besides pumice can trap buoyant air bubbles. Even a lowly piece of aluminum foil can trap enough air to stay afloat. Here's how.

Material Required

- Small square of aluminum foil;
- Small container;
- Water.

Procedure

Crumple a small square of aluminum foil into a loose ball. The ball should be about the size of a large marble. Make sure there are many nooks and crannies within the material in which air bubbles can fit.

Drop the crumbled foil ball into a container of water. The ball should float on the surface of the water. Now shake the aluminum ball beneath the surface of the water. Keep shaking it until there are no visible air bubbles on the foil's surface. Release the ball and this time the ball should sink.

This is What Happens

The overall density of the foil ball determines its buoyancy. When it was first crumpled, the ball contained many air spaces and bubble-trapping nooks and crannies. These extra air spaces decreased the overall density of the ball and made it less dense than water. As a result, the ball floated.

When the bubbles were shaken away, the overall density of the ball increased. With this greater density, the foil ball sank.

166. Balloon Rocket

Do unbalanced forces produce motion? This is what we are going to find out.

Material Required

- Yardstick (meter stick);
- Drinking straw;
- Scissors;
- String;
- Balloon, 9 in. (23 cm);
- 2 chairs;
- Masking tape.

Procedure

Cut a 4 in. (10 cm) piece from the drinking straw. Cut about 3½ ft. (4.5 m) of string. Thread the end of the string through the straw piece.

Position the chairs about 4 ft. (4 m) apart. Tie the string to the backs of the chairs. Make the string as tight as possible. Inflate the balloon and twist the open end. Move the straw to one end of the string. Tape the inflated balloon to the straw. Release the balloon.

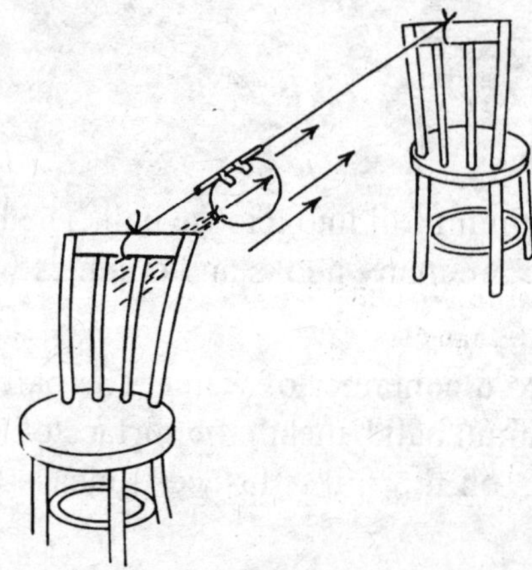

This is What Happens

The straw with the attached balloon jets across the string. The movement stops at the end of the string or when the balloon totally deflates.

Science Behind It

Newton's Law of Action and Reaction states that when an object is pushed, it pushes back. When the balloon was opened, the walls of the balloon pushed the air out. When the balloon pushed against the air, the air pushed back and the balloon moved forward, dragging the straw with it. The string and straw keep the balloon rocket on a straight course.

167. Cartesian Diver

A pump spray perfume bottle and a Cartesian diver are two very different items, yet they share some basic physics. When you push down on the pump, you produce a force. This force pushes on the perfume stored within the bottle. The pressurized liquid rises through a tube and emerges as a mist through the spray nozzle.

But what has squeezing a pump spray bottle have to do with a submarine-type toy that dives on command? Find out.

Material Required

- Glass medicine dropper with bulb;
- or 2-litre plastic soft-drink container with cap;
- Water.

Procedure

Fill the soft drink container to the brim with water. Draw water into the dropper until the glass pipette is about 1/3 filled with water. Put the dropper into the container. If needed, add more water to the container. Tightly screw on the cap.

Grasp the container. Slightly increase the pressure of your grip. Observe the appearance of the trapped air bubble within the dropper. Keep squeezing and you'll soon discover why this device was given the name diver.

This is What Happens

When you squeezed the container, you produced a force. This force acted upon the water in the container. Since the water could not be easily compressed, the force was transmitted "through" the water. Within the diver, the force pressed upon the air bubble. Air, unlike water, is easily compressed. Under the stress of this force, the bubble shrank in size. As it shrank, its buoyant force (B.F.) decreased. As soon as the B.F. became less than its weight, the diver sank.

168. Change in Energy

What happens to energy after it is used? This is what we are going to see by performing this easy experiment.

Material Required

- Ruler;
- String, 24 in. (60 cm);
- Duct tape;
- Book;
- 2 small rubber balls of equal size.

Procedure

Cut a 24 in. (60 cm) piece of string. Insert one end of the ruler into a book. Tie the centre of the string around the end of the ruler. Use very small pieces of tape to attach the hanging ends of the string to the balls. The tape sticks best if the balls are clean and oil-free. The strings on the balls must be the same length. Pull the balls away from each other and release them.

This is What Happens

The balls continue to hit and bounce away from each other until they finally stop moving.

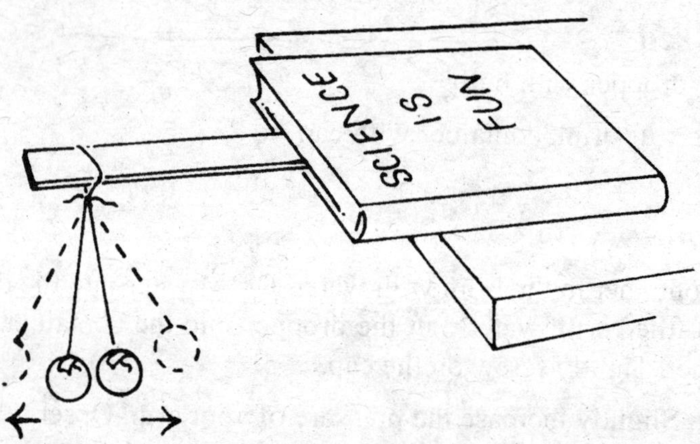

Science Behind It

The *Law of Conservation of Energy* states that energy is never lost or created. The energy to raise the balls to a higher position came from you. Releasing the balls allowed the energy due to their height, or potential energy, to be changed to energy of motion, or kinetic energy. The balls exchanged energy with each other when they collided. They also received a push in the opposite direction, resulting in backward motion after contact. The balls stop moving when their energy is changed to other forms of energy such as heat and sound.

169. Mini Submarine

A submarine's ability to sink and surface is set by the balance of forces between the submarine's weight and its buoyancy. Although the weight of the hull remains mostly the same, the buoyant force is increased or reduced.

Submarines have compartments called ballast tanks. Ballast tanks can be opened and closed. In order to dive, the tanks open. As they fill with water, the submarine becomes less buoyant and sinks. In order to surface, pressurized air is injected into the tanks. This air blast "blows out" water to make the submarine more buoyant.

Material Required

- Film canister with lid;
- String;
- Tape;
- 1/4 teaspoon baking powder;
- Manual hand drill;
- Large bowl;
- Water;
- Spoon;
- 1 ounce sinker (fishing weight).

Procedure

Using a hand drill to carefully place six small holes (each about 1/8 inch in diameter) in the bottom of a film canister.

Tie a string to a 1-ounce fishing sinker. Use tape to secure the sinker's string to the bottom of the film canister. Place about 1/4 teaspoon of baking powder into the canister. Seal the canister with its lid.

Place the canister and sinker in a large bowl of water. As the canister fills with water, it becomes less buoyant. When it begins to sink, release the canister. Keep watching. In a few moments, the evolving gas should turn things around.

This is What Happens

When the canister first filled with water, it sunk. Within moments, however, the water that entered the canister mixed with the baking powder to produce carbon dioxide gas. As the gas was generated, it rose to the top of the canister. There the gas bubbles took the place of water. Water that had previously occupied this space was forced downward out of the canister.

This action of replacing one substance by another is called displacement. When enough water had been displaced, the B.F. was strong enough to bring the canister to the surface.

170. Mass and Motion

Does mass has any effect on kinetic energy (energy of motion) on any object? This is what we are going to check out here.

Material Required

- Table;
- Small pail;
- Masking tape;
- Paper;
- Pencil;
- Wooden block;
- String, 1 yd. (1 m);
- Scissors;
- Clay.

Procedure

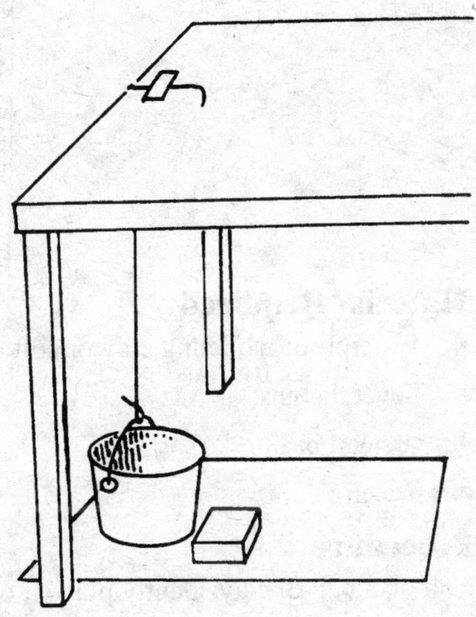

Attach the string to the handle of the pail. Secure the opposite end of the string to the edge of the table. The string needs to be long enough to allow the pail to swing about 1 in. (2.5 cm) above the floor. Place the paper on the floor under the hanging pail. Position the wooden block on the floor in front of the hanging pail. Pull the pail back and allow it to swing into the block. Mark the position that the block has moved to on the paper.

Again position the wooden block on the floor in front of the hanging pail. Place large pieces of clay in the pail to increase its weight. Pull the pail back to the same position as before, then allow the pail to swing into the wooden block. Mark the position of the block on the paper.

This is What Happens

The wooden block moves farther when struck by the heavier pail.

Science Behind It

The pail was raised to the same height each time, keeping the speed constant. The mass of the pail increased when the clay was added. Kinetic energy (energy of motion) is increased as the mass of a moving object is increased. The heavier pail had more energy when it struck the wooden block, and thus it pushed the block farther along the paper.

171. Floating Concrete Ships

Do you know in some of the old ships the hull is made of concrete? Have you ever wondered why ships made of a material heavier than water can float? What keeps them from sinking to the bottom?

Material Required

- Waterproof modeling clay or soft clay used by florists;
- Several coins;
- A small basin;
- Water.

Procedure

Roll up a ball of clay. Gently place it on the water's surface. Release it. What happens?

Now shape the lump of clay into a small boat. For a real challenge, you'll want to build a clay vessel that can support the weight of several coins.

This is What Happens

So far, floating has been explained in terms of two concepts: 1) the balance between weight and buoyant force; 2) the differences in density.

Here's a third concept that can be applied to floating: displacement.

When you sit in a bath tub, the level of water rises. It increases because your body now occupies a space that was previously filled with water. When you sat down, you pushed away (or displaced) an equal volume of water. When you displace water, you generate an upward force that is equal to the weight of the water you "pushed out."

A ship's hull also displaces water. It is designed to displace enough water to keep the entire boat (hull, decks, stacks, etc.) afloat.

172. Paper Helicopter

The purpose of this experiment is to determine how weight affects the rotation speed of a paper helicopter.

Material Required

- 3 paper clips;
- Pencil;
- Notebook paper;
- Scissors;
- Ruler.

Procedure

Fold and cut one sheet of paper in half lengthwise. Take one of the halves and fold it in half lengthwise. Use a ruler to draw a triangle on one edge of the paper. The base of the triangle will be 1 in. (3 cm) long and one side will be between the 4 in. and 6 in. (9 cm and 14 cm) marks on the ruler. See the diagram.

Cut out the triangle. Cut through both layers of the paper. Open the paper and cut up the centre fold to the point indicated on the diagram. This forms the two wings. Fold the tabs towards the centre and attach a paper clip to the bottom. Fold the wings in opposite directions.

Hold the helicopter above your head and drop it. Add different numbers of paper clips one at a time and drop the plane after each addition. Observe the speed of rotation after each paper clip is added.

This is What Happens

The rotation speed increases as the weight increases, but a point is reached where additional weight pulls down with such force that the wings move upward and the plane falls like any falling object.

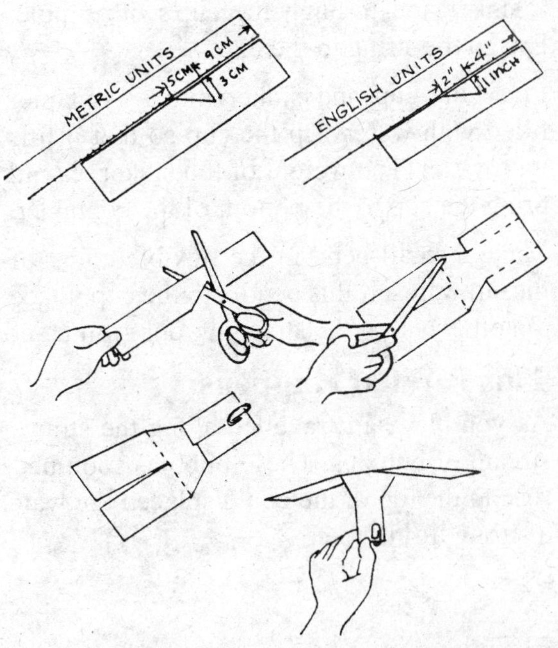

Science Behind It

As the paper falls, air rushes out from under the wings in all directions. The air hits against the body of the craft, causing it to rotate. Increasing the weight by adding paper clips causes the helicopter to fall faster, and the mount of air hitting the craft's body increases. This increases in air movement under the wings increases the rotation speed.

173. Balloon Left

In the Classical novel by Jules Verne, The Mysterious Island, several people are trapped on an isolated island. To escape, they raise a sunken wreck by stuffing a large hot air balloon into its hull and pumping it full of air. As the balloon inflates, it displaces water that had filled the sunken hull. When enough water is displaced, the ship rises to the surface.

Material Required

- Paper cup;
- 1 ounce lead sinker;
- Water-filled container;
- String;
- Pencil;
- Flexible straw.

Procedure

Using a pencil, poke two small holes beneath the rim of the paper cup. The holes should be positioned at opposite sides of the cup.

Cut a length of string about 8 inches long. Insert the string into one of the holes and tie it. Then pass the string through the "eye" of a sinker and through the cup's other hole. Fasten the string to the cup.

Place the cup and sinker into a container filled with water. Tip the cup so that it fills with water. The weight of the sinker should bring the cup to the bottom of the container.

Make a small bend in the flexible collar of the straw. Insert this bend into the cup. Place the other end of the straw in your mouth and blow. What happens to the cup?

This is What Happens

As you blew, air travelled along the straw. When it exited the straw, it floated upwards as a stream of bubbles. These bubbles soon became trapped within the inverted cup. The air that rose to the top of the cup displaced the water. Eventually, enough air filled the cup to produce a strong lifting force.

174. Right or Left Rotation

How wing position affects the direction of a paper helicopter's rotation? This is what we are going to try out.

Material Required

- Paper helicopter from experiment Paper Helicopter.

Procedure

Hold the helicopter above your head and drop it. Observe the direction that the helicopter spins. Bend the wings in the opposite direction and again drop the helicopter from above your head.

This is What Happens

The helicopter spins counterclockwise when the right wing is bent towards you and turns clockwise when the wings are reversed.

Science Behind It

Air rushes out from under each wing in all directions as the helicopter falls. The air hits against the body of the craft, pushing it forward. Both halves of the body are being pushed in a forward direction, resulting in a rotation about a central point. The diagrams indicate direction of movement.

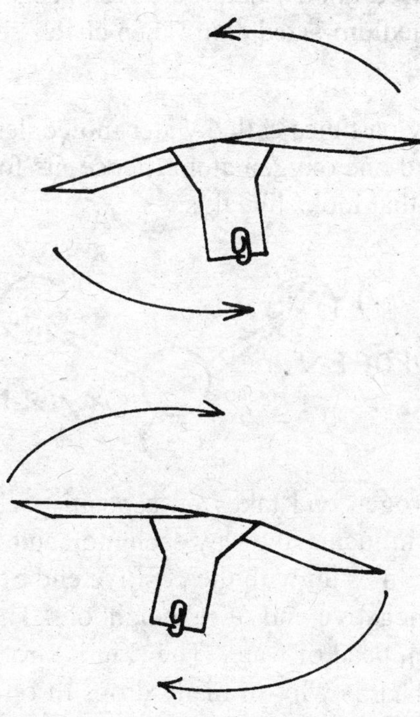

175. Coin Pool

Have you ever examined a drop of water that had fallen onto a sheet of wax paper? If so, you know that the drop wasn't flat. Instead, it formed a bead. This round shape results from tiny, unseen forces that exist in the smallest particles of water.

Material Required

- Small size coin;
- Medium size coin;
- Water dropper;
- Water.

Procedure

Make a guess. How many drops of water can fit on the head of a coin without spilling over the edge? Write down your guess.

Now use a dropper to carefully add water to the head of a small coin. Count each drop. Try not to splash the drops or move the coin. When you have finished, compare the actual number of drops with your guess. How close were you?

Let's make another guess. Based upon what you have observed, predict the number of drops that can fit on the head of a medium-sized coin. Then check your answer.

This is What Happens

Water is made up of very tiny particles called water molecules. Each water molecule is made up of two hydrogen atoms and one oxygen atom (hence, its formula H_2O). This molecule can be represented by something that looks like this:

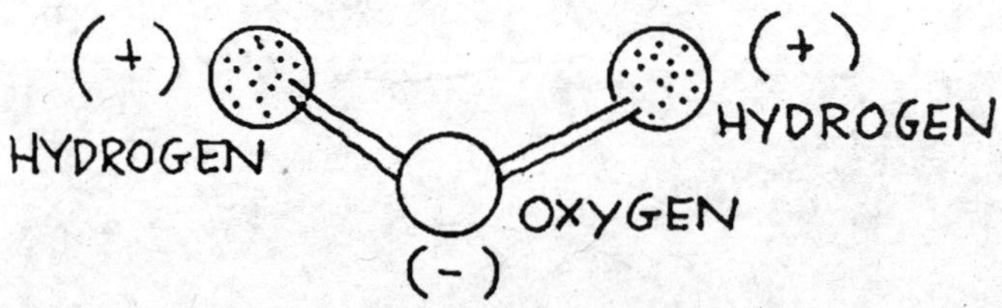

But, there's more. Each hydrogen end takes on a positive charge. Each oxygen takes on a negative charge. The charges influence the way each molecule lines up with its neighbour. The negative end of one molecule lines up with the positive end of its neighbour. The positive end of another lines up with the negative end of its neighbour. This attraction results in a type of "skin" that holds together each bead of water. The skin is strong enough to stretch and hold in additional water molecules. That is why so many drops fit on the head of the coin. Break the attraction between neighbouring molecules and the bead flattens by spilling water over the coin's rim.

176. Friction Force

Has the texture of a surface anything to do with motion? In this experiment, we will determine the effects.

Material Required

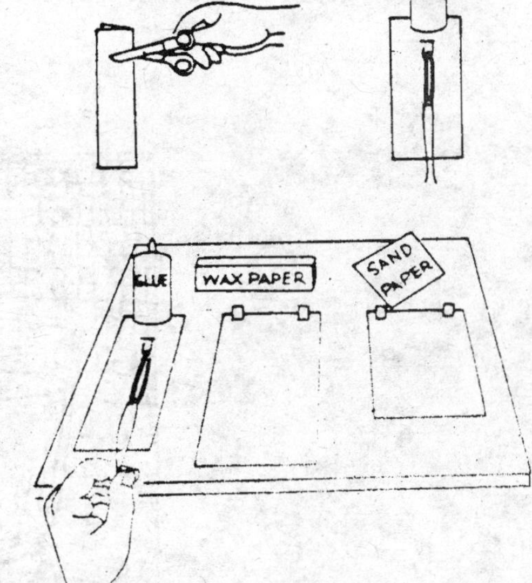

- Poster board;
- Ruler;
- Paper clip;
- Rubber band;
- Scissors;
- Bottle of glue, 8 oz. (236 ml);
- String;
- Pen;
- Wax paper;
- Sandpaper.

Procedure

Cut a 5 in. x 10 in. (12 cm x 25 cm) card from the poster board. Fold the card and cut a slit about 1/2 in. (1 cm) long 2 in. (5 cm) from the end of the card. Place the paper clip in the slit and slip the rubber band on the paper clip. Cut a 10 in. (25 cm) piece of string and loop it through the rubber band. Place the paper on a table. Position the bottle of glue at the end of the card. Gently pull on the string to straighten the rubber band.

Mark the card at the end of the rubber band and label this mark START. Pull on the string until the card begins to move. Note how much the rubber band stretches. Tape a sheet of wax paper and a sheet of sandpaper to the table. Move the card with the glue bottle across the wax paper and sandpaper by pulling on the string. Observe how much the rubber band stretches each time.

This is What Happens

The rubber band stretched the least when the card was placed on the wax paper and stretched the most on the sandpaper.

Science Behind It

The weight of the glue bottle pushes the card down against the surface it sits on. The card is more easily pulled across the wax paper than across the sandpaper because of friction. Friction is a force that pushes against a moving object, causing the object to stop moving. Frictional force increases with the roughness of the surfaces moving against each other. The surface of the wax paper is smoother than that of the sandpaper or table, and thus applies less fractional force to the card.

177. Quarter in the Cup

Now that you understand water tension it's time to play "Quarter in the Cup."

To win, all you have to do is guess the numbers of quarters that can be placed into this cup without water spilling over the rim.

Material Required

- Clear plastic cup or glass;
- Stack of quarters (25 Paisa coins);
- Water.

Procedure

Fill the cup or glass to the brim with water. Examine the cup. Guess how many quarters can be placed into it without spilling water over the rim. Write this guess down. Now see how close you are to your prediction by gently dropping quarters into the cup.

This is What Happens

Like the previous experiment, this one is based upon the attraction between neighbouring water molecules. As quarters are lowered into the cup, they displace (or "push out") an equal volume of water. The water level rises.

Although the water rises above the rim, it doesn't spill over. It's that attraction between the opposite charges of oxygen and hydrogen that keeps the "skin" together. The skin, however, does have limits. When the surface becomes too curved, the balance is destroyed and water spills out over the rim.

178. Changing Energy

The purpose of this experiments is to demonstrate the effect that height has on the energy of a moving object.

Material Required

- Book;
- Pencil;
- Ruler with a centre groove;
- Paper cup, 8 oz. (236 ml);
- Scissors;
- Marble.

Procedure

Cut a 1 in. (2.5 cm) square section from the top of the paper cup. Place the cup over the ruler. The end of the ruler should touch the back side of the cup. Raise the opposite end of the ruler and rest it on the pencil. Place the marble in the centre groove of the ruler at the ruler's highest end. Release the marble and observe the cup.

Raise the end of the ruler and rest it on the edge of the book. Again position the marble in the groove at the ruler's highest end. Release the marble and observe the cup.

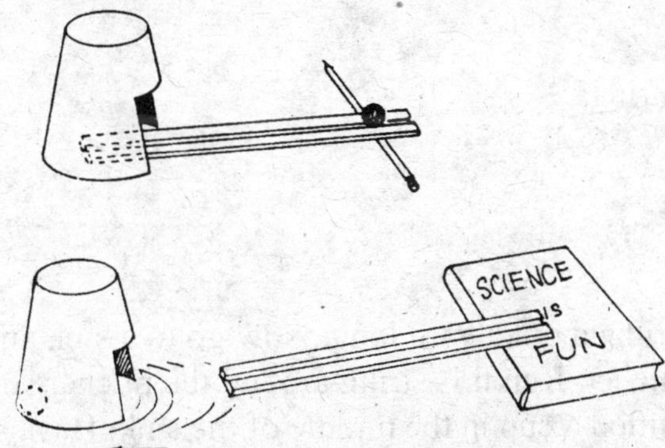

This is What Happens

The cup moves when the marble strikes it. The cup moved an additional distance when the ruler rested on the book.

Science Behind It

Objects at rest have potential energy. The higher the object sits above the ground, the greater is its potential energy. When objects fall or roll down an incline, their potential energy changes into kinetic energy—energy of motion. Increasing the height from which the marble rolled gave it more energy, causing it to strike the cup with more force.

179. Chain Attraction

If you spill something in the kitchen, it'll be a mess. If you spill something in the laboratory, it can be deadly!

Scientists use tricks to transfer liquids. These tricks prevent splashing and make sure that liquids pour into beakers instead of running down onto sneakers.

Material Required

- Water-filled pitcher;
- Cup;
- 1-yard-long kite string;
- A pair of scissors;
- Sink.

Procedure

Cut a length of kite string about a yard long. Now go to a sink and wet the string. Tie one end to the handle of the pitcher. Run this string around the pitcher so that it keeps in contact with the pouring spout. Position a cup in the middle of the sink. Have a friend hold the other end of the string over the centre of the cup. Make sure that the string is pulled tight. Then gently pour out the water. What happens?

This is What Happens

Water flowing over the spout runs down the string. As long as the string is kept taut, water will flow along its length. At the bottom of the string, the water passes into the cup (without a mess).

Remember how water molecules beaded up on a sheet of wax paper? That was due to surface tension. Well, an attraction between molecules is at it again. This time, it keeps the water molecules attached to each other in a stream-like chain.

180. Slide or Roll

The idea behind this experiment is to find out how different surfaces affect friction.

Material Required

- String, 24 in. (60 cm);
- Rubber band;
- 2 large books;
- 10 round marking pens;
- Ruler.

Procedure

Stack the books on a table. Tie a string around the bottom book. Attach the string to the rubber band. Move the stack of books by pulling on the rubber band. Measure how far the rubber band stretches. Place the 10 marking pens under the stack of books. Move the books by pulling on the rubber band. Observe how far the rubber band stretches.

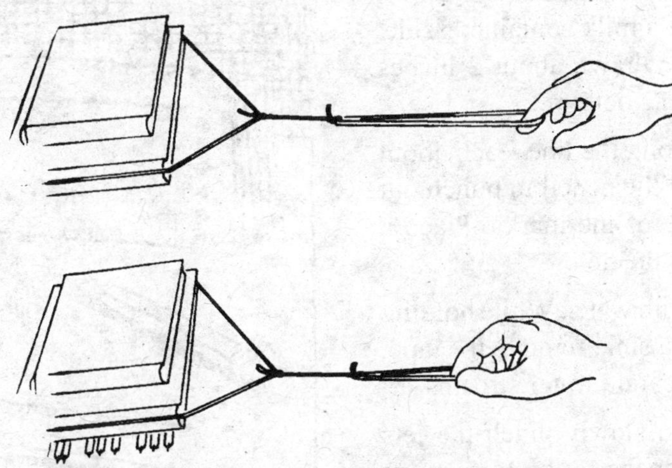

This is What Happens

The rubber band stretches more when the bottom book sits flat against the table than when it is placed on the pens.

Science Behind It

Friction is a force that tries to stop motion. The flat surface of the book slides across the table and the round pens roll across the table. Things that roll cause less friction than things that slide. Thus, there is less friction between pens and the table than between the book and the table.

181. United Streams

In rainy season, you have seen water droplets appear to follow a choreographer's directions when flowing down a window. These directions seem to be laid out in a pattern of droplets. When a droplet encounters another droplet, they join. Quickly the two fall and encounter other droplets. Eventually, they drip out of sight!

Material Required

- Milk container or large drink box;
- Sharp pencil;
- Ruler;
- Tape;
- Water;
- Sink.

Procedure

Use a ruler to draw a horizontal line across the bottom edge of a milk container's side. The line should be drawn about 2 inches above the container's bottom.

Make six marks along the line, each about 1/2 inch apart. Use the pencil to punch out a hole at each one of the marks. Place a strip of tape across the holes.

Fill the container with water. While holding the container over a sink, remove the tape. Observe the six separate water streams.

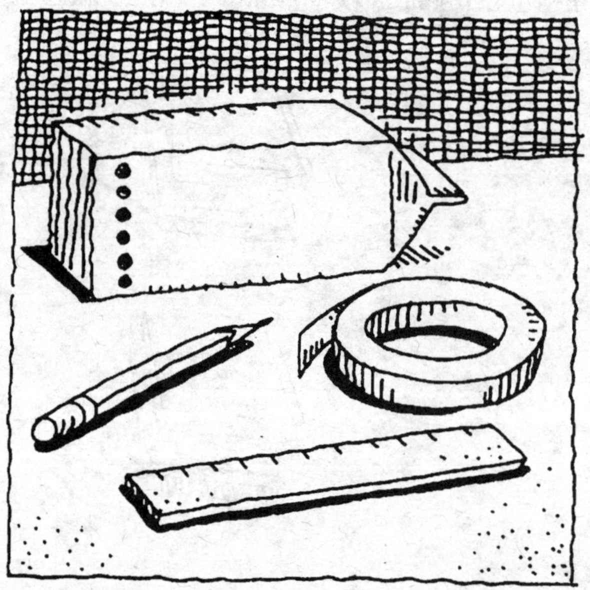

Use your fingers to slowly pinch the jets together. What happens to the streams as they are being pinched? How long will they remain together? How can you make them flow separately again?

This is What Happens

As you probably guessed, the streams' behaviour is due to the attraction between water molecules. At first, the individual water streams are too far apart to affect each other's path. However, when they are pinched, the streams are brought together. Their force of attraction is strong enough to prevent the streams from separating. However, if a hand physically separates them, they will go back to flowing as six individual streams.

182. Simple Hover Craft

The purpose of this experiment is to demonstrate how motion is affected by friction.

Material Required

- Cardboard;
- Pencil;
- Empty thread spool;
- Balloon, 9 in. (23 cm);
- Ruler;
- Scissors;
- Glue;
- Notebook paper.

Procedure

Cut a 4 in. (10 cm) square from the cardboard. Use the pencil to punch a hole through the centre of the cardboard square. Glue the empty thread spool over the hole in the cardboard. Be sure the hole in the spool lines up with the hole in the cardboard. Place a bead of glue around the base of the spool. Cut and glue a circle of paper over the top end of the thread spool. Allow the glue to dry for several hours.

Use a pencil to punch a hole in the paper circle to line up with the hole in the spool. Place the cardboard on a smooth surface such as a table. Push the cardboard and observe its motion.

Inflate the balloon and twist the end. Stretch the opening of the balloon over the top of the spool. Untwist the balloon and give the cardboard a little push and observe its motion.

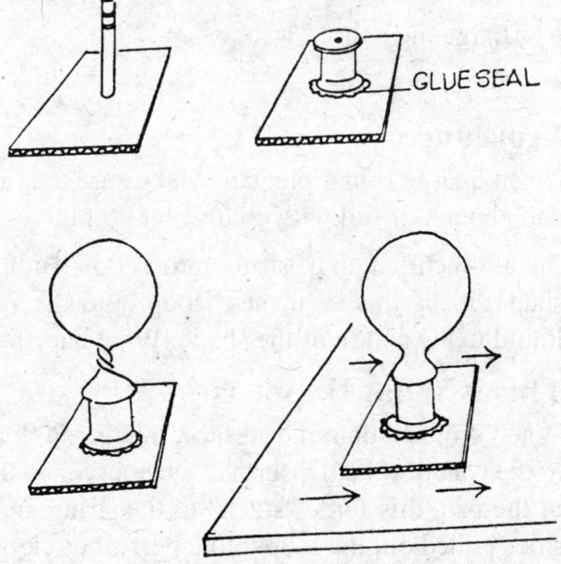

This is What Happens

The hover craft moves very little without the balloon, but easily glides across the table when the air flows through the spool.

Science Behind It

The air flowing from the balloon through the holes forms a thin layer of air between the cardboard and table. This air layer reduces friction (a force that tries to stop movement), allowing the hover craft to quickly move across the table.

183. Dancing Loop

Have you ever wanted to choreograph a dance? Well, here's your chance. Your dancer, however, won't be made of flesh and bones. Instead, it will be made of cotton.

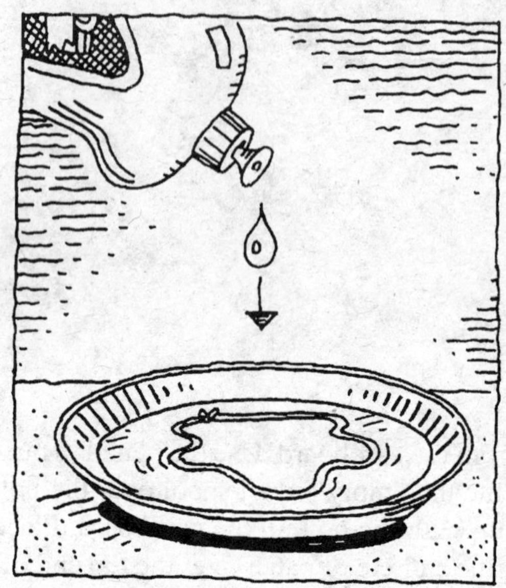

Material Required

- 6-inch-long thread;
- Dishwashing liquid;
- Large pie pan;
- Water.

Procedure

Clean a large round pie pan. Make sure that all soap has been rinsed from its surface. Fill the pan about half full with cold water. Let the water stand until it is perfectly still.

Tie a 6-inch length of string into a loop. Bunch up the thread so that the loop has an irregular shape. Place this "squished" loop onto the surface of the water. Add a drop of dishwashing liquid to the centre of the shape. What happens?

This is What Happens

When dropped onto the surface, the thread floated atop a layer of surface tension. Adding soap to the water broke the surface tension. Since the surface tension was still active around the rim of the pan, this force "stuck" to the string and pulled it out from the centre. Equal tugs to all sides pulled out the loop into a perfect circle.

184. Making a Paddle Boat

By making this paddle boat, it will be much easier for you to understand Newton's Law of Action and Reaction.

Material Required

- Cardboard;
- Rubber band;
- Scissors;
- Container of water at least 4 in. (10 cm) deep;
- Ruler.

Procedure

Cut a 4 in. (10 cm) square from the cardboard. Shape the boat by cutting one side into a point and cutting out a 2 in. (5 cm) square from the opposite end. Cut a paddle from the cardboard. Make it 1 in. x 2 in. (2.5 cm x 5 cm). Loop the rubber band over the ends of the boat. Insert the paddle between the sides of the rubber band. Turn the cardboard paddle towards you to wind the rubber band. Place the boat in the container of water and release the paddle. Observe the direction of motion.

Wind the rubber band in the opposite direction by turning the paddle away from you. Place the boat in the water and release the paddle. Observe the direction of motion.

This is What Happens

The boat moves forward with the first trial and backwards when the paddle turned in the opposite direction.

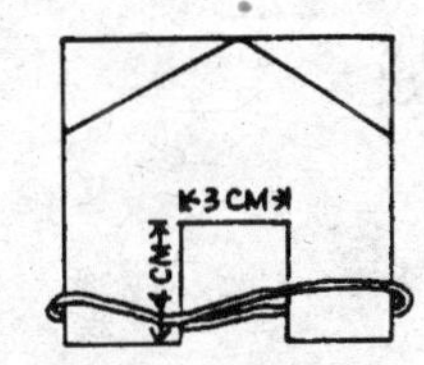

Science Behind It

Newton's Law of Action and Reaction states that when an object is pushed, it pushes back with an equal and opposite force. Winding the paddle caused it to turn and hit against the water. When the paddle pushed against the water, the water pushed back and the boat moved. The boat moved in the opposite direction to the paddle, changing direction when the paddle direction changed.

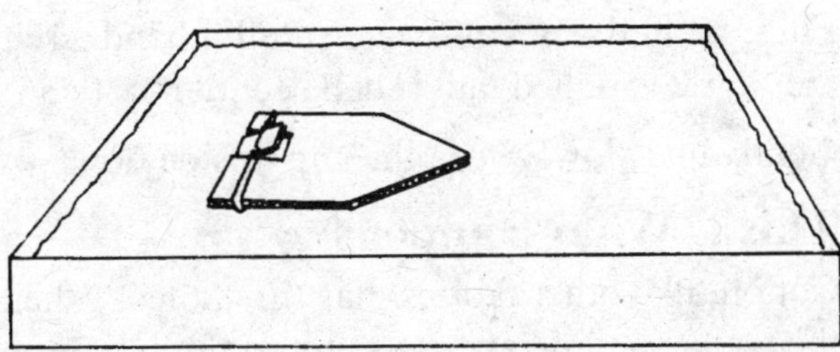

185. Surface Tension

It's raining, it's pouring... You know how it goes. This experiment, however, isn't about covering heads and going to bed. It's about surface tension and how it relates to umbrellas.

Material Required

- Small jar;
- Cotton handkerchief;
- Rubber band;
- Water;
- Sink.

Procedure

Fill a jar halfway with water. Pull a handkerchief across the mouth of the jar. Make sure that the fabric is pulled taut. Then secure it with a rubber band.

Gently and slowly, turn the cup upside down over a sink. What happens to the water?

This is What Happens

Cotton, like other fabrics, has tiny holes. When the jar was turned upside down, the weight of the water should have caused it to flow through the fabric's holes. It didn't.

Although you couldn't see it, the holes were patched by water molecules. Surface tension formed waterproof plugs across the fabric's holes. This skin was strong enough to prevent water from flowing through the porous fabric.

186. Air Brakes

This fine experiment demonstrates the effect of friction on motion.

Material Required

- String;
- 2 paper clips;
- Empty styrofoam thread spool;
- Ruler;
- Cellophane tape;
- Stiff paper (index card will work);
- Scissors;
- Knitting needle.

Procedure

Ask an adult to cut four slits at right angles in the thread spool. Cut four 3 in. x 1 1/2 in. (7.5 cm x 4 cm) cards from the stiff paper. Cut a 16 in. (40 cm) length of string and tape it to the side of the spool. Attach two paper clips to the end of the string. Put the knitting needle through the centre of the spool. Wind the string around the spool. Hold the knitting needle and observe the speed of the unwinding string.

Insert the four paper cards in the slits in the spool. Wind the strings around the spool. Hold the knitting needle and observe the speed of the unwinding string.

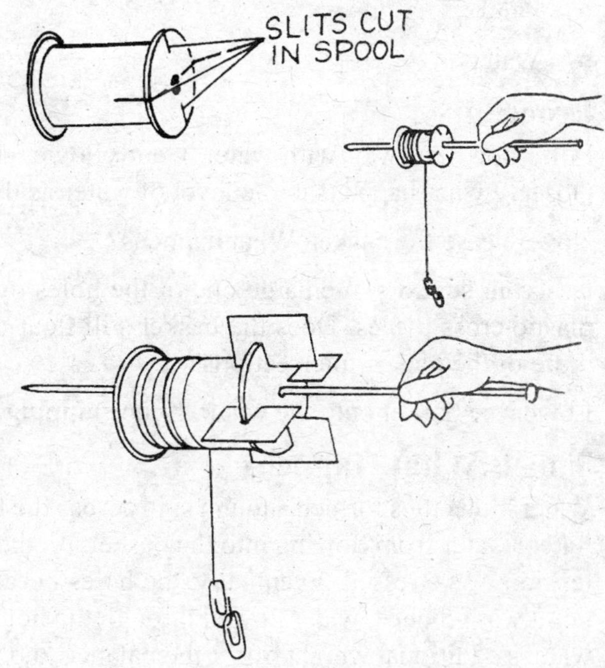

This is What Happens

The spool turns more slowly when the paper cards are in place.

Science Behind It

Gravity causes the paper clips to fall and pull the attached string with them. As the string unwinds, it spins the spool. Friction is the resistance to motion. There is friction between the paper cards and the molecules of gas in the air. Air pushes against the paper cards as the spool spins, reducing the speed of the turning spool.

187. Plugging the Holes

Here is one experiment that will easily explain surface tension.

Material Required

- Plastic screen food basket;
- Bowl;
- Pair of scissors;
- Water.

Procedure

Fill a bowl halfway with water. Gently lower the food basket onto the surface of the water. Observe what happens to the level of water as the basket is lowered.

Now release the basket. What happens?

Use your scissors to enlarge one of the holes by making four small snips into the connecting plastic cross pieces. Does the basket still float? If so, enlarge the hole by snipping away one more of the plastic intersections.

Lower the basket onto the water. Keep snipping off pieces until the basket can no longer float.

This is What Happens

Water molecules formed a tough skin across the holes of this basket. The skin was so tough that it kept water from flowing into the basket. As the hole size was increased, however, the surface tension was stressed. Eventually, the holes became too large to keep out the water. When that point was reached, water flowed through the hole and entered the basket. Within the basket, the water's additional weight offset the balance and the basket sank.

188. Rolling Distance

The reason of this easy experiment is to determine the effect of weight on the energy of a moving object.

Material Required

- 2 books;
- 1 small round jar;
- 1 large round jar;
- Yardstick (meter stick).

Procedure

Set the edge of one book on the second book to form an incline. Place the small jar at the top of the incline. Measure the distance from where the jar stops to the end of the book. Allow the large jar to roll down the incline and measure the distance it rolls.

This is What Happens

The larger, heavier jar rolls farther than the lighter jar.

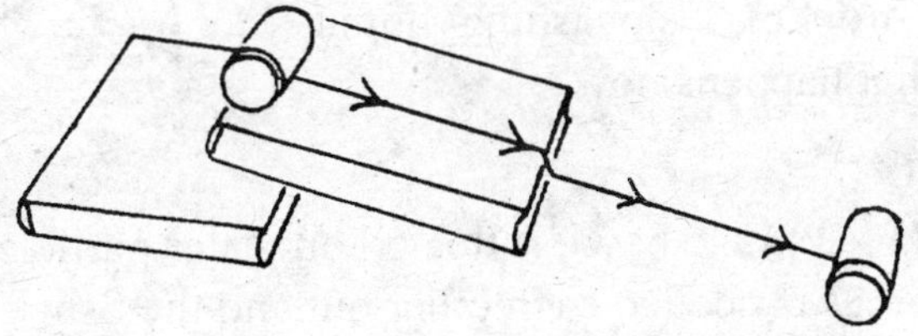

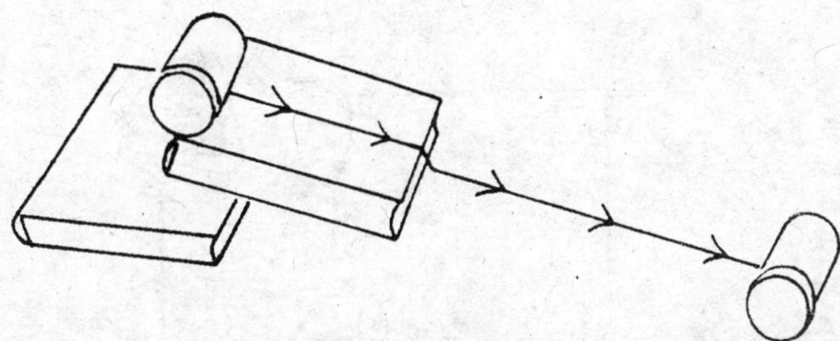

Science Behind It

Since the friction of the air, book, and floor is the same for both experiments, it will not be considered. The major difference is the weight of the two jars. As the weight of the rolling object increases, its energy increases. The energy of the separate rolling jars is calculated by multiplying each weight by the height of the incline. Keeping the height the same—constant—makes the weight of each jar responsible for the change in the rolling distance.

189. Soap Fuelled Fish

In good old days, "motorized" fish were sold as cheap toys. These fish were constructed from paper. According to the instructions, all you had to do was add a drop of the right solution and off went the fish. But what was the right solution?

Material Required

- Small bowl;
- Pair of scissors;
- Water;
- Heavy stock paper;
- Dishwashing liquid.

Procedure

Fill a small bowl with water. Use a pair of scissors to cut out the fish pattern shown below.

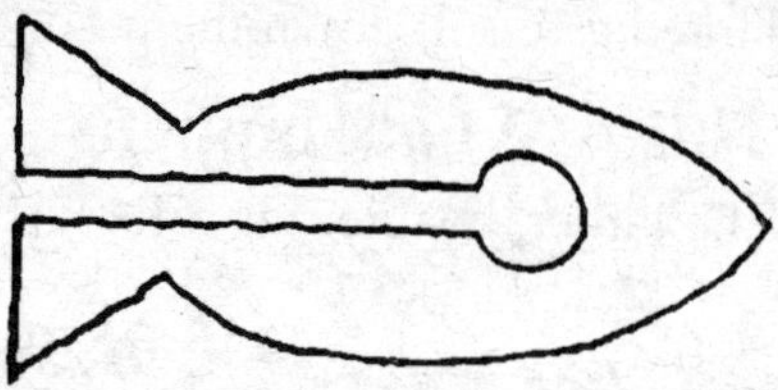

Place the pattern in the centre of the bowl. Observe its movement. Now place a drop of dishwashing liquid within the centre circle. What happens now?

This is What Happens

When the paper fish was placed in the bowl, it floated on water particles that pulled equally in all directions. These tiny forces cancelled each other out and the fish remained still.

When a drop of dishwashing liquid was added, the balance of surface tension was destroyed. As the soap flowed out the back of the fish, it broke up the surface tension. Since the pulling force at the front was still intact, the fish moved forward.

190. Wobbling Water Motion

In this experiment, the results will be just opposite to the experiment, you have performed earlier (Rolling Distance). Be prepared for the unexpected results.

Material Required

- 2 books;
- 1 jar with lid, 1 qt. (1 litre);
- Yardstick (meter stick).

Procedure

Place the edge of one book on top of the second book to form an incline. (See diagram.) Close the empty jar and place it at the top of the incline. Release the jar. Measure the distance the jar rolls from the end of the book.

Fill the jar three-quarters full with water. Secure the lid and place the water-filled jar at the top of the incline. Release the jar and record the distance the jar rolls.

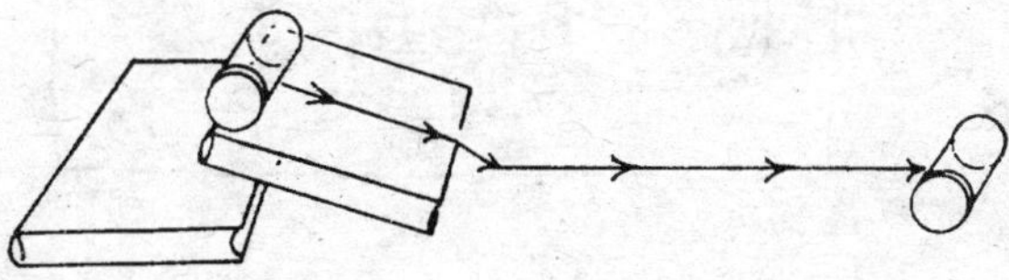

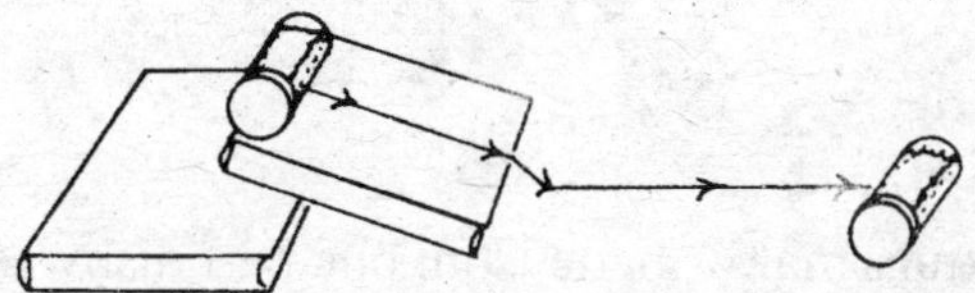

This is What Happens

The empty jar rolls farther.

Science Behind It

The friction from the air, book and floor is the same during both experiments; thus, they do not have to be considered. In experiment Rolling Distance, it was discovered that heavier objects roll farther. The water makes the jar heavier, but the water hits against the inside of the jar as it rolls down the incline, increasing the friction inside the jar. It takes more energy to move this jar with the wobbling water inside.

191. Bubbles Machine

Have you ever played with a magnet and a paper clip? If so, you know that as the magnet and clip are separated, the magnetic attraction decreases.

Now imagine the tiny forces between neighbouring water particles. While they are side by side, their attraction is strong. But if we separate the particles, the attraction is lessened.

Material Required

- Cup;
- Dishwashing liquid;
- Measuring cup;
- Pipe cleaner;
- Water.

Procedure

Fill a cup halfway with a mixture that is made from one part dishwashing liquid and nine parts water. Gently, but thoroughly, mix this solution.

To form a bubble wand, bend one end of a pipe cleaner into a large, loop. Dip the bubble wand into your solution. Remove the wand and a film of bubble solution will stretch across the loop opening. As you gently blow into this loop, bubbles will emerge from the opposite side.

This is What Happens

When dishwashing liquid is added to water, the liquid breaks down into very tiny particles. These particles of dishwashing liquid spread out and get between neighbouring water molecules. There, they interfere with the attraction between water molecules. This allows the water to spread apart into a thin film. Without the interference of detergent, the intense attraction between the water particles would "snap" the shape shut and not allow the skin to be drawn into a film. When this film surrounds a pocket of air, it becomes a bubble.

192. Marble Movements

The purpose of this experiment is to determine how the speed of a rotating object is affected by its distance from the centre.

Material Required

- Record player;
- Round cake pan;
- Marble;
- 1 sheet of construction paper, any colour;
- Masking tape, 1 roll;
- Scissors.

Procedure

Cut a circle from the paper to fit inside the pan. Place the roll of tape in the centre of the record player turntable. Place the cake pan with its paper lining on top of the tape roll. Turn the record player on low speed and adjust the tape roll and pan to place them exactly in the centre of the turntable. Increase the speed to high. Place the marble in different places in the spinning pan. Observe the speed at which the marble moves. Place the marble exactly in the centre of the rotating pan.

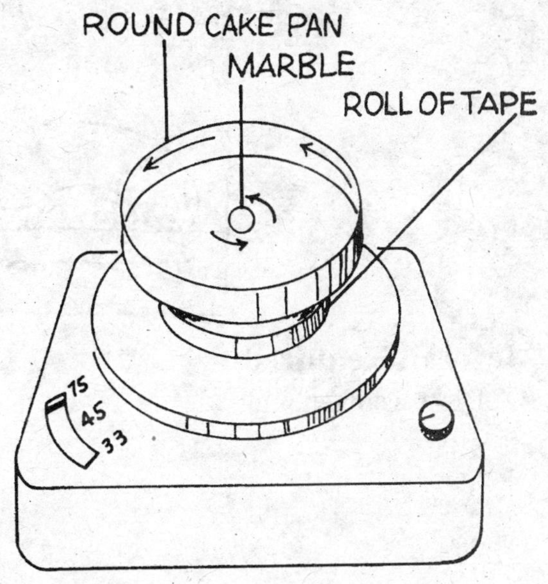

This is What Happens

The marble moves quickly when placed near the side of the pan. The marble stays in place when placed in the centre of the pan.

Science Behind It

The rotating motion of the pan moves the marble outward. The speed of the marble depends on how close it is to the rotating centre. The closer to the centre, the more slowly the marble moves outward, until a point is reached where the marble's outward speed equals zero; that point is the exact centre.

193. Bubble in Bubble

Here's a bubbly challenge. Can you place a bubble within a bubble? Perhaps. But can you place a third bubble within second...and a fourth within the third?

Material Required

- Dishwashing liquid;
- Cup;
- Water;
- Straw;
- Sink.

Procedure

Make a bubble solution using one part dishwashing liquid and nine parts water. Dip the end of a straw into the bubble solution.

Wet the surface of a sink with a layer of bubble solution. Place the wetted end against the wet sink surface. Gently blow into the dry end of the straw. Once the dome bubble has been made, withdraw the straw. The bubble seals itself and remains stable.

Now rewet the straw and inject it into the bubble. Once the tip touches the sink surface, blow your next bubble.

194. Pepper Race

In this experiment, we will make black pepper run across a bowl of water with soap as its power source.

Material Required

- Black pepper;
- Toothpicks;
- Bowl full of water, 2 qt. (2 litre);
- liquid detergent;
- Saucer.

Procedure

Sprinkle the pepper over the surface of the water in the bowl. Pour a few drops of liquid detergent into the saucer and dip the end of the toothpick into the detergent. Insert the wet end of the toothpick in the centre of the pepper.

Note: All the soap must be washed out of the bowl and fresh water used before the experiment can be repeated.

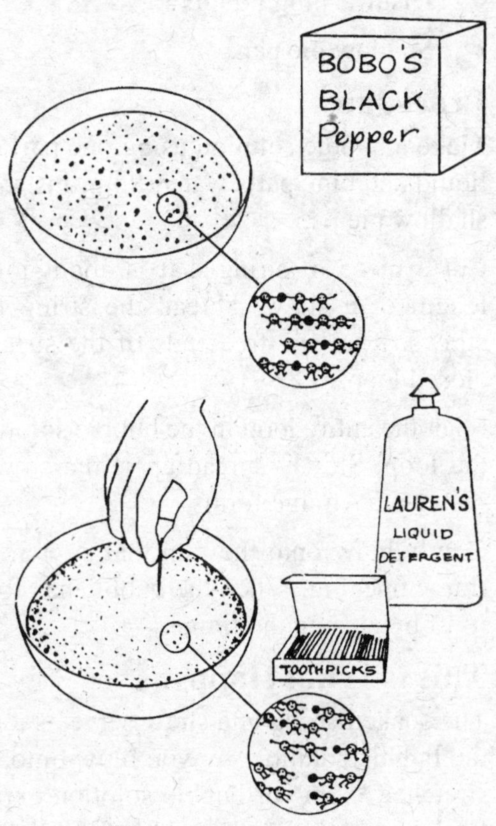

This is What Happens

The pepper breaks in the centre and runs towards the sides of the bowl.

Science Behind It

Each pepper speck is waging a tug-of-war battle. While the water is clean, the surface water molecules pull on the pepper specks with equal force in all directions. Placing the soap in the centre weakened the pull of the water molecules in that area, and the stronger, clean-water molecules pulled the pepper specks across the surface of the water towards the side of the bowl.

195. Making Bubble Frame

Did you know that your hand can be a great bubble maker? Just make an "okay" sign with your thumb and index finger. Soak this "loop" into a bubble solution. Then blow into it gently and steadily.

Neat, huh? Well, here's another type of bubble maker. This one is called a bubble frame.

Material Required

- Kite string;
- Pair of scissors;
- Two straws;
- Cup;
- Dishwashing liquid;
- Shallow pie pan.

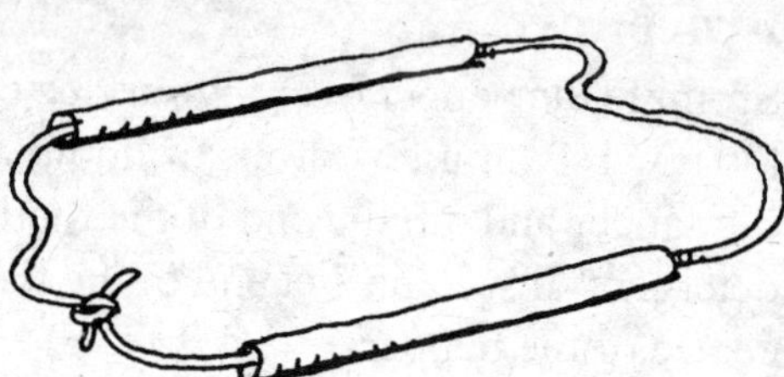

Procedure

Make a bubble solution using one part dishwashing liquid and nine parts water. Pour this solution into a shallow pie pan.

Cut a piece of string that is about four times the length of a straw. Thread the string through two straws. Knot the free ends of the string forming a closed loop.

Soak the entire loop in the bubble solution. Remove the loop. Slowly spread apart the straws so that a rectangular frame forms.

Gently blow onto the film that stretches across the frame opening. When the bubble has formed, pinch it off by closing the frame.

This is What Happens

The soaked string and straw serve as a reservoir for the bubble solution. As you blow onto the film, the stretched "skin" of bubble solution expands. When the frame is closed, the bubble film sticks back onto itself and pinches off from the plastic straw and cotton string.

196. Tricky Balloons

How the size of a balloon affects the air pressure inside it? This is what we will try to figure out.

Material Required

- 2 round balloons, 9 in. (23 cm);
- Plastic tubing, 12 in. (30 cm) 1ong;
- Rubber band;
- Clothespin.

Procedure

Insert one end of the plastic tubing inside the mouth of a balloon. Use the rubber band to tightly secure the balloon's mouth to the end of the tube. Blow through the tube to inflate the balloon. Bend the tube in half and secure it with the clothespin. This should keep the air inside the balloon. Inflate the unattached balloon to about half the size of the first balloon. Twist the neck of the balloon to prevent air loss, then insert the free end of the tube into the balloon's mouth. Hold the mouth of the smaller balloon tightly around the tube and remove the clothespin. Observe the change in the size of the balloons.

This is What Happens

The smaller balloon deflates.

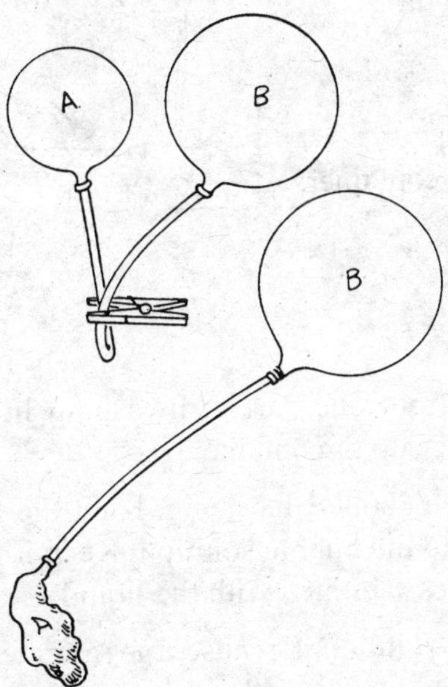

Science Behind It

If you could divide the larger balloon in half, the amount of air in one of the halves would be the same amount as in the single small balloon, but the size of the half would be much smaller than the single balloon. When comparing equal amounts of air molecules, the smaller balloon is more stretched and pushes its air into the larger balloon.

197. Big Bubble Maker

If you want to make the biggest possible bubble, there are several things to keep in mind. First, the biggest bubbles form in humid surroundings. So try to avoid the dry air of winter.

Also, direct sunlight is bad for bubbles. Try making them in the shade. And, of course, you'll need a special bubble maker such as the one described below.

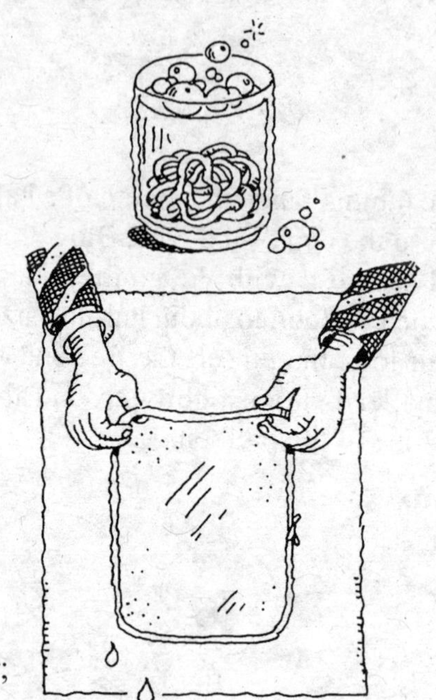

Material Required

- 3-foot-long cotton rope ;
- A large plastic cup or container;
- Pair of scissors;
- Dishwashing liquid.

Procedure

Make a bubble solution from one part dishwashing liquid and nine parts water. Pour this solution into a large plastic cup or container.

Cut a length of cotton rope about 3 feet long. Knot the free ends of the rope to form a large loop. Immerse the loop into the bubble solution. Keep it in the solution for several minutes so that the entire rope becomes saturated with the liquid.

Remove the rope from the solution. Because it is saturated, it will sag. Position your hands on opposite sides of this closed loop. Gently and steadily spread the loop apart. The bubble film should span the opening.

Slowly walk with the loop extended. The passing air will enter the loop and inflate the bubble. When the bubble has grown to a large enough size, close the loop so that the bubble gets pinched off.

198. Fly Away

The purpose of this experiment is to demonstrate movement of air due to unequal pressure.

Material Required

- Narrow-necked soda bottle;
- 1 sheet of notebook paper.

Procedure

Lay the empty soda bottle on its side. Tear off one-fourth of the paper sheet and squeeze it into a ball. Place the wad of paper just inside the opening of the bottle's neck. Try to blow the paper wad into the bottle with your breath.

This is What Happens

The paper wad flies out of the bottle.

Science Behind It

Before you blow into the bottle, the amount of air inside and outside the bottle are the same. The air pressure inside the bottle is increased by the extra air blown into the bottle. This extra air is pushed out of the opening, and the moving air swishes the paper wad out into the air.

199. Pouring Air

Imagine pouring a glass of water. As you tilt the container, the water rises to the brim. Soon it overflows the lip of the glass. As it flows over the edge, the water forms a steady downward stream. Downward, as in going down towards the floor. But did you know that some fluids pour upwards?

Material Required

- Two clear drinking glasses;
- Aquarium or large fish bowl;
- Water.

Procedure

Fill an aquarium tank 1/2 full with water. Place one glass in the tank so that it fills completely with water. Then turn it upside down.

Take a second glass, hold it upside down, and push it into the water. What happens to the glass? Does it fill with water?

Position this air-filled glass near the bottom of the tank and directly beneath the water-filled glass. Tilt the air-filled glass so that a bubble of air escapes. Where does the air go?

Continue tilting the glass so that its air contents flow upward and into the other glass. What happens to the water-filled glass as air enters it? Can you explain this observation?

This is What Happens

When the air-filled glass was tilted, the air escaped as rising bubbles. Instead of reaching the surface, the stream of rising bubbles were "intercepted" by the opening of the water-filled glass. There, the rising air became trapped and filled this second glass.

200. Different Waves

How sound and light travel? This is what we are going to find out by using your toy slinky (spring type toy).

Material Required

- Slinky, about 4 in. (10 cm) long;
- Helper.

Procedure

Have your helper hold one end of the Slinky and stretch it about four times its length while you hold the other end. Shake your end of the Slinky up and down several times. Notice how the Slinky moves.

Place the Slinky on the floor and stretch it as far as possible. Reach as far as you can down the stretched spring and gather the coils towards you, then quickly release them. Notice how the Slinky moves.

This is What Happens

Shaking the Slinky up and down makes it look like water waves. Compressing the coils sends a back-and forth pulsing movement down the spring.

Science Behind It

The compression-and-expansion type of motion is called a longitudinal wave. Sound is produced by a vibrating material that moves back and forth. As the material moves outward, it pushes the air in front of it together, producing a compressed area of air molecules. When, the material swings back, it leaves an expanded area. This back-and-forth motion produces compressed and expanded area in the air through which the energy is being transferred. This is how sound vibrations move through the air.

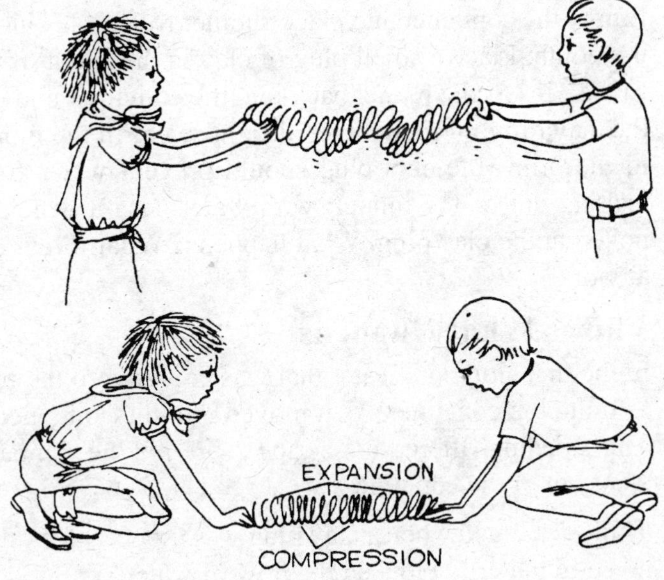

The up-and-down movement of the Slinky represents a transverse wave. These are like water waves in that they have high and low parts called crests and troughs. Light travels in transverse waves.

201. Down the Drain

Use a can opener to punch one hole near the rim of a small liquid-filled can. Turn the can on its side, and see what happens. Not much. Although there's a hole in the container, the liquid doesn't seem to flow out. Something is in the way!

Material Required

- Funnel;
- Clay;
- Cup;
- Pencil;
- Clear plastic beverage container.

Procedure

Place a roll of clay around the mouth of a plastic container. Insert a funnel in the mouth and press the clay firmly against the spout to form an airtight seal. Slowly pour water from a cup into the funnel. What happens? Can you describe the movement of two fluids, even the one you can't see?

Empty the container and place another roll of clay along its mouth. Place a small plug of clay in the spout of the funnel. Insert the funnel back into the container and use the clay roll to form another airtight seal. Fill the funnel to the brim (the clay plug should prevent water from leaking out of the funnel) with water. Use a pencil to poke out the clay plug. What happens? What forces are at work?

This is What Happens

In the first situation, water that was poured into the funnel swirled around its inner surface and fell into the container. As it moved into the container, it did not fully block the spout opening. This opening offered an escape route for the trapped air. Water went into the container. Air went out of the container.

Things changed when a clay plug was used to block the spout. When the plug was knocked out, no solid material blocked the flow of water.

There was, however, a container full of trapped air which kept the water out. Since the water filled the entire spout opening, there was no escape path for the air.

202. Always Straight

The purpose of this experiment is to show that light travels in a straight line.

Material Required

- Cardboard;
- Flashlight;
- Scissors;
- Modeling clay;
- Ruler;
- Index card.

Procedure

Cut three 6 in. (15 cm) squares from the cardboard. Cut 1 in. (2.5 cm) square notches from the centre of one edge of each of the three cardboard squares. Use the clay to position the cardboard about 4 in. (10 cm) apart with the notches aligned in a straight line. Lay the flashlight behind the notch of one of the end cards. Use clay to position the index card screen at the end opposite the flashlight. Darken the room and observe any light pattern on the paper screen. Move the cardboard so that the notches are not in a straight line. Observe any light pattern on the paper screen.

This is What Happens

A circle of light appears on the screen only when the notches are in a straight line with each other.

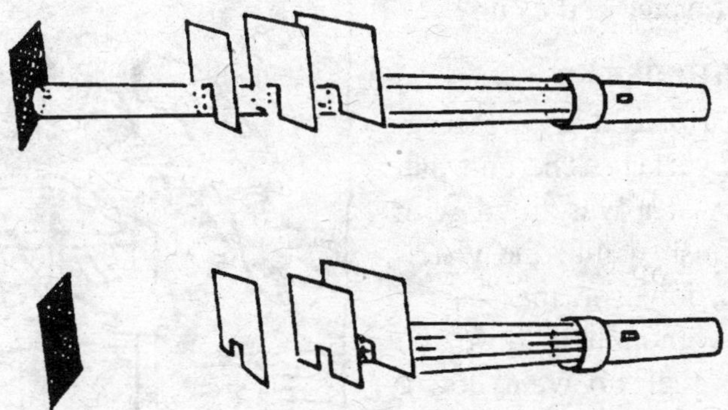

Science Behind It

Light travels in a straight line. When the notches were in line, the light rays were able to pass through the openings, but the rays were blocked by the cardboard when the notches were out of line.

203. Underwater Boat

Think of an underwater boat and most likely two images come to mind: submarines and sunken wrecks. In this experiment, you'll construct a third type of submerged vessel. Unlike the other two, this one will actually "float" beneath the water's surface. Interested?

Material Required

- Clear plastic cup;
- Large transparent container, such as a fish bowl or aquarium tank;
- Water;
- Aluminum foil;
- Toothpick;
- Paper;
- Clay.

Procedure

Build a small boat out of aluminum foil, paper, and other construction materials. Float this boat on the surface of the water-filled fish bowl.

Place an empty and clear cup over the boat and gently push down. What happens to the boat as the cup is pushed into the water? Does the boat continue to float? Why? What happens to the level of water in the cup?

This is What Happens

When your boat was placed in water, it floated on the liquid's surface. The cup that was placed over the water was filled with air. As this cup was pushed into the water, its trapped air pushed down on the surface below. This push was enough to drop the water's surface. So as the cup went down, so did the boat which still remained at the surface of this trapped water pocket (although this surface was now underwater).

204. Light and Vision

Let's test the movement of light through different materials in this simple experiment.

Material Required

- Cardboard;
- Wax paper;
- Plastic sandwich bag.

Procedure

One by one, hold the plastic, wax paper, and cardboard pieces in front of one of your eyes. Look through the materials and note any differences in how objects in the room appear.

This is What Happens

There is little or no change in appearance when things are observed through the plastic. The wax paper makes objects look dull and frosty, while nothing can be seen through the cardboard.

Science Behind It

In order for you to see anything, light must be reflected from the object you are looking at to your eyes. The clear plastic is an example of a transparent material. Transparent means that light moves straight through the material and allows you to see objects as they are. Translucent ma-terials, like wax paper, change the direction of the light that passes through. This change in direction results in objects looking dull, frosty, and sometimes distorted. Cardboard is an opaque material—no light can pass through. Without light passing through to your eye, nothing on the opposite side of opaque materials can be seen.

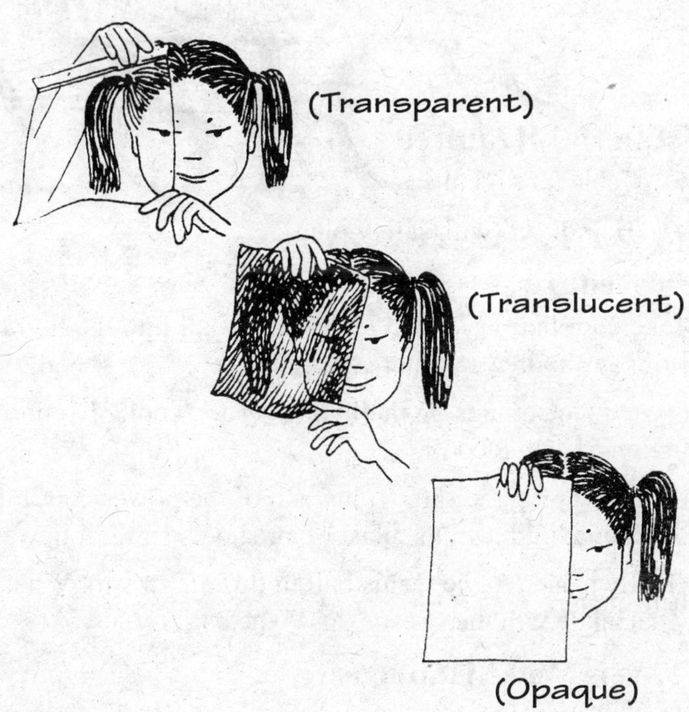

205. Floating Balls in Air

In this experiment, we will try to float Table Tennis ball in the air with the help of hair dryer.

Material Required

- Table tennis ball;
- Hair dryer (with safety plug).

Procedure

Take your hair dryer and table tennis ball into the living room. Perform this experiment away from any source of water.

Switch the settings on the hair dryer to "cool." Do not use any of the dryer's hot or warm air options. Keep it cool!

Carefully plug the safety plug end of the power cord into an electrical outlet. Turn on the hair dryer and hold it so the blast of air shoots straight upwards.

Gently place a table tennis ball in this air stream. What happens? How far is the ball shot? At what angle will the stream "lose" the ball?

This is What Happens

The ball floats on the jet of fast moving air whose motion produces a region of low pressure. Since the surrounding air has a slightly greater pressure, it pushes inward on the stream from all directions. This inward force is strong enough to "trap" the table tennis ball within the air jet. If, however, the stream is tilted too much, the pressure difference cannot overcome the pull of gravity and the ball falls from the stream.

206. Polarized Light

The aim of this experiment is to show how polarized light moves.

Material Required

- Two pairs of polarized sunglasses.
- **Note:** Be sure you are using polarized glasses.

Procedure

Put one pair of glasses on. Observe how objects around you appear. Hold the second pair of glasses in front of your eyes. Slowly rotate the pair of glasses you are holding so that one of the lens turns in front of your right eye. Observe the quality of vision as you turn the glasses.

This is What Happens

One pair of sunglasses seems to cut down glare and change the shade of objects. During the rotation of the second pair of glasses, the objects seen through the right eye got darker until finally nothing could be seen through the two lenses.

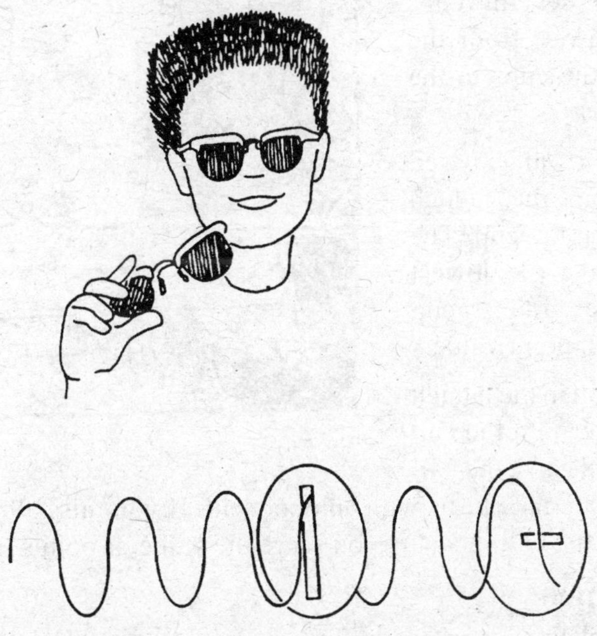

Science Behind It

A polarized lens has an endless number of parallel slits. Light waves moving in the same direction as the slits are allowed to pass through. Light not moving in the same direction as the slits in the lens is blocked and cannot pass through.

207. H-Tech Skeeball Game

Skeeball is a game in which you must land a wooden ball into a hoop. You start by rolling the ball down a ramp. At the end of the ramp, the ball becomes airborne. If your aim was accurate, the ball lands within a small hoop. If you miss your target completely, the ball rolls back down the slope with little or no scored points.

In this experiment, you'll get to assemble a similar game. You won't, however, roll the ball with your arm. Instead, take a deep breath as you read the rules.

Material Required

- Table tennis ball;
- A marker;
- Several small cups;
- Tape.

Procedure

On a desktop, place several cups at various distances away from the "launch" cup. Secure the cups to the desktop with tape.

Give each cup a point value based upon its distance from the launch cup. Write this value on the cup. The closest cup should have the lowest point value while the farthest cup should have the greatest point value.

Place the table tennis ball in the launch cup. Direct a blast of air into the cup at an angle so that it will carry the ball towards a target cup. Alternate turns with an opponent. If you miss all cups, you don't receive any points. If you land the ball in a cup, you score the value of points listed on the cup's side.

This is What Happens

It's that low pressure phenomenon again. The stream of moving air creates a region of low pressure. As the ball pops out of the launch cup, it gets "trapped" within the low pressure flow and was kept in place by the surrounding high pressure air. As gravity takes its toll, the ball drops—ideally into the cup with the highest point value!

208. Rainbow on Water Surface

In rainy season, you may have seen rainbow in the sky. It's a wonderful sight to see one. But in this experiment, we will try to create one on the surface of water. Believe me! You can do it, just follow the instructions.

Material Required

- Bowl of water, 1 qt. (1 litre);
- Bottle of clear fingernail polish.

Procedure

Place the bowl of water on a table away from direct lighting. Hold the brush from the polish bottle over the bowl of water and allow one drop of liquid polish to fall into the bowl. Watch the surface of the water. Move your head so that you see the surface from different angles.

This is What Happens

A rainbow of colours is seen in the thin layer of fingernail polish on the surface of the water.

Science Behind It

The nail polish forms a thin film across the water. When light rays strike the film, part of each ray is reflected from the surface. Part of the ray goes through and is reflected off the bottom of the film. If the reflected rays overlap as they leave the film, colours are seen. The timing has to be just right for the reflected rays from the surface and bottom of the film to meet as they leave the film; if this does not happen, areas without colours are seen. This rainbow of colours is called a spectrum.

209. Air Pressure

With the help of your old plastic darts with suction cups and playing cards you can perform this experiment.

Material Required

- Two suction cup darts;
- Playing card;
- Thumbtack;
- Tape;
- Water.

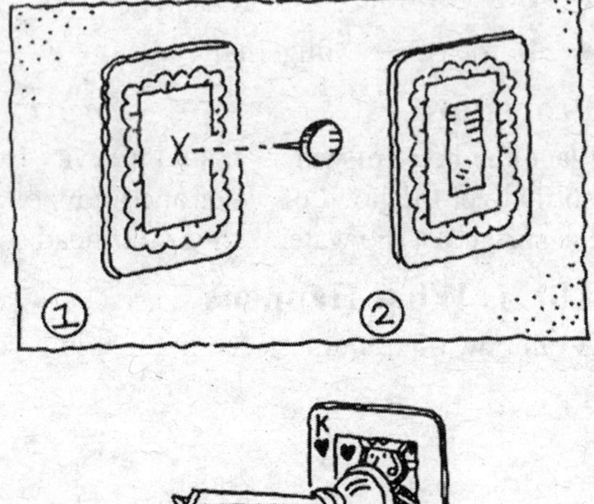

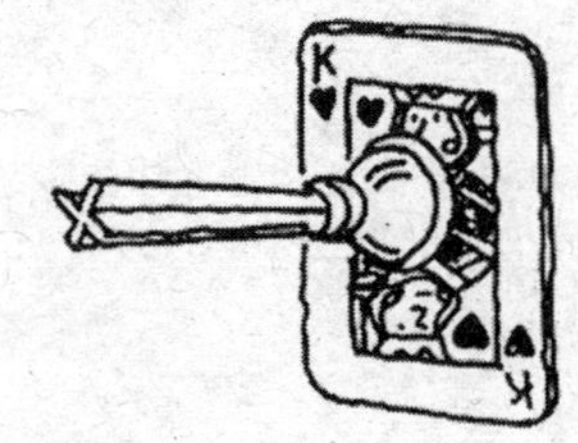

Procedure

Make a hole through the centre of a playing card with a thumbtack. Use a small piece of tape to cover one side of this hole. Push a suction cup dart against the playing card so that it covers the untaped side of the hole. What happens?

Moisten the "lip" of the suction cup. Try sticking the cup against the card again. Does wetting the suction cup affect the way it sticks?

While the cup is stuck to the card, remove the tape. What happens to the suction cup if the tape is peeled away. Can you explain your observations?

This is What Happens

The air that surrounds us pushes out with a force we call air pressure. The amount of air pressure is related to the concentration of air particles. The greater the concentration of particles, the greater the pressure. The fewer the particles per space, the lower the pressure.

When the suction cup is pushed against the card, air is forced out of the cup. Since air leaves this space, there are fewer particles trapped within the cup. As the cup "unsquishes," it produces a slight drop in the concentration of air particles. Hence, there is a lower air pressure inside the cup.

Outside the cup, however, things have stayed mostly the same. Since the surrounding air has a greater concentration of air particles, it has a higher pressure. This pressure pushes in the cup and keeps it from moving off its attached surface.

When the tape was pulled off the card, air rushed into the low pressure chamber. As the concentration of air particles equalized, the force keeping the cup stuck to the card dwindled.

210. Water Prism

You may have used prism in your science classes to separate light into its separate colours. In this experiment, we will make a prim out of water.

Material Required

- Flashlight;
- Heavy paper;
- Masking tape;
- Scissors;
- Sheet of white paper;
- Chair;
- Glass of water;
- Helper.

Procedure

Cut a circle from the heavy paper to cover the end of the flashlight. Cut a very thin slit across the circle, stopping about 1½ in. (1 cm) from the edge. Tape the paper circle to the front of the flashlight. Place the glass of water on the edge of the chair.

Have your helper hold the white paper near the floor at the edge of the chair. Darken the room and hold the flashlight at an angle to the surface of the water. Change the angle of the flashlight and ask your helper to vary the position of the white paper. Look for colours on the white paper.

This is What Happens

A spectrum of colours is seen on the paper.

Science Behind It

White light contains all of the spectrum colours—red, orange, yellow, green, blue, indigo, and violet. Light can be separated into the spectrum colours by passing it through different substances such as water or glass. Light must be refracted —bent or spread out— in order to be split into a spectrum. Scientists use this property of refraction to determine the density and composition of materials through which light can pass.

211. Magical Cup

We live our lives surrounded by an ocean of air. Although we aren't aware of it, this air has weight. The weight pushes down, in, out, and up on everything. Even though we can't feel this air pressure, we can certainly observe its effects.

Material Required

- Paper;
- Thumbtack;
- Piece of cardboard;
- Water;
- Paper cup;
- Sink.

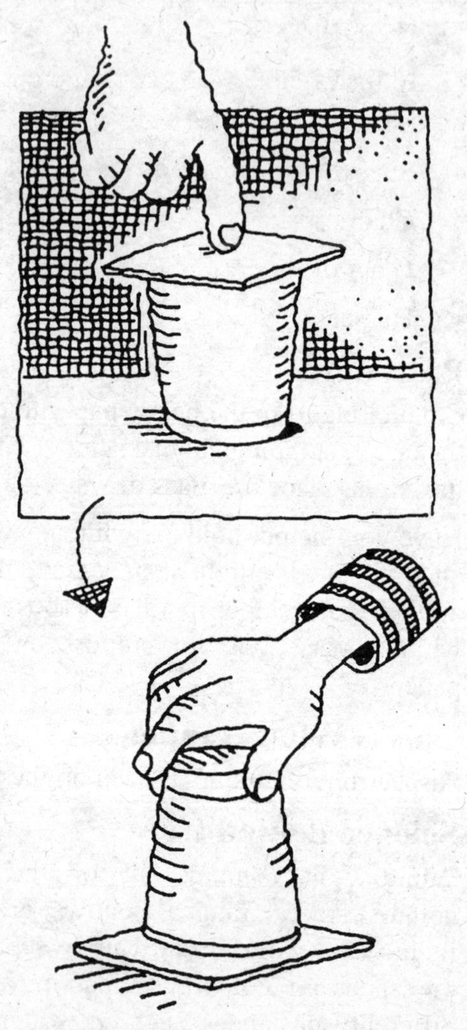

Procedure

Fill a paper cup to the brim with water. Then carefully place a small square-shaped piece of cardboard over the mouth of the cup.

Place one finger in the centre of the cardboard. Then, while holding the cup over a sink, slowly turn the cup over. Remove your finger and the cardboard remains in place.

Gently punch a thumbtack hole into the bottom of the cup that is now facing upwards. What happens?

This is What Happens

Under the weight of the water, the cardboard bottom sags slightly. This "drop" produces a slight drop in concentration of air, which is trapped in the cup. This drop produces a region of low pressure. Since the outside air remains at the same concentration, it retains a relatively greater pressure than the air inside the cup. This pressure difference is enough to keep the card in place even though it is supporting a cup full of water!

When a hole was punched into the cup, air rushed in to equalize the pressure. As the pressure balanced, the card fell away under the water's weight.

212. Blender

This simple experiment demonstrates how light waves blend to produce white light.

Material Required

- Poster board;
- Scissors;
- Pencil;
- Ruler;
- Marking pens, red, orange, yellow, blue, green, and violet.

Procedure

Cut a 4 in. (10 cm) circle from the poster board. Divide the circle into 6 equal sections. Colour the sections in this order: red, orange, yellow, blue, green, and violet. Punch the pencil through the centre of the circle, leaving about half of the pencil on each side. Place the point of the pencil on a flat surface and quickly spin the top of the pencil. Observe the colour of the spinning circle.

This is What Happens

The spinning circle looks grey.

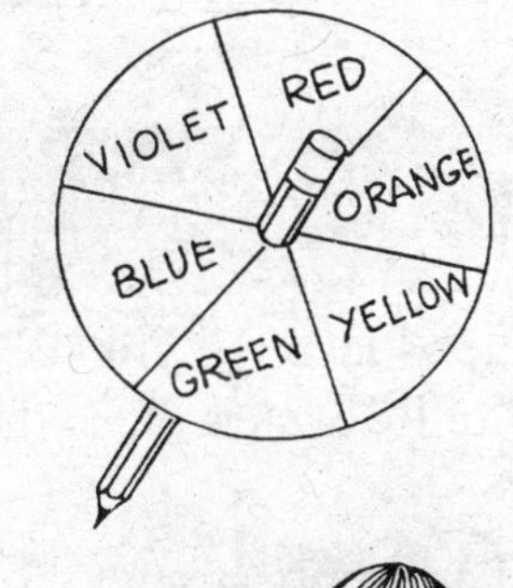

Science Behind It

White light is made up of all the spectrum colours—red, orange, yellow, green, blue, indigo, and violet. As the coloured circle rotates, the different colours are perceived by your eyes. The images of the colours are retained in your mind long enough to cause an overlapping of the messages. Your own mind blends the colours together. If the colour indigo—a dark bluish colour—were included, you would see a twirling white card because all of the spectrum colours blended together produce white light.

213. The Straw Dropper

It's cold outside and you have to take some couch syrup. Opps! You have misplaced medicine dropper that was needed. If so, don't worry. Here's how you can use a straw to make a precision liquid delivery system.

Material Required

- Straw;
- Cup filled with water;
- Sink.

Procedure

Insert a straw into a cup filled with water. Place your finger over the straw's open end, making an airtight seal. Then lift the straw out of the water. Place the straw over a sink and release the seal. What happens?

Now place your finger over one end of an empty straw to form an airtight seal. While keeping this seal, insert the straw into a cup filled with water. Wait several seconds. Remove the straw and place it over a sink. Release the seal. What happens now?

This is What Happens

When you immersed the open-ended straw into the water, it filled with liquid. By sealing the top of the straw, you formed an airtight chamber within the straw. When the straw was lifted out of the water, the liquid within the straw sagged downwards (you might have observed the curved drop at the straw's bottom). The slight drop was enough to create a pressure difference between the chamber and the outside surrounding air. That air kept pressing against the liquid in the straw to prevent it from spilling out.

When you immersed the closed-ended straw into the water, it did not fill. By sealing the end, you created an enclosed air space. This chamber of air exerted a pressure that kept water from entering up the straw. Since no water could get in, the straw remained dry.

214. Red Skin Filter

In this experiment, we will determine how filters affect light.

Material Required

- Flashlight.

Procedure

Darken a room and hold the flashlight under your hand. Move the light around behind your fingers and palm. Observe any light that passes through.

This is What Happens

Parts of your hand appear rosy or red in colour.

Science Behind It

Your flesh and skin act like filters. A filter is any material that absorbs some of the colours in light and allows others to pass through. Red filters absorb all colours except red which passes through. Your skin takes on a rosy colour because the red blood under the skin acts like a red filter—red light passes through and other colours are stopped.

215. Tower of Water

Although you can't see it, the surface of a glass of water is under plenty of pressure. The pressure is formed from the weight of a column of air that stretches upwards for over 20 miles! Although the water's flat and calm surface may not offer clues to this unseen force, here's a way for you to observe its effects.

Material Required

- Clear plastic container;
- Plastic soft drink bottle;
- Water.

Procedure

Fill the bottle with water. Position it above the empty container. Invert the bottle so that the water flows into the container. Keep the mouth of the bottle positioned about 1 inch from the bottom of the container. What happens when the water level rises to the level of the bottle's mouth?

Raise the bottle to 2 inches from the bottom. What happens now?

Refill the bottle until the container is $^1/_3$ full of water. Cover the bottle's mouth to form a water-tight seal. Invert the bottle and place its mouth below the water line. Remove your hand. What happens to the level of water within the container? Can you explain your observation?

This is What Happens

When the bottle is turned upside down, the weight of the water causes this fluid to flow downwards. It keeps flowing until the water level rises to the level of the bottle's mouth. At this point, the level of the rising water "shuts off" the bottle's flow.

The weight of the surrounding air pushes down on the surface of the rising water. This downward push is transferred in all directions through the particles of water. The force that is transferred in the upward direction resists the water spilling out of the bottle.

Since the atmospheric force counteracts the weight of the water, it stops the downward flow and prevents additional water from spilling out of the bottle.

Although the tower of water supported in this activity was only a few inches tall, standard air pressure can keep up a column of water that is higher than a 3-story building!!!

216. False Image

How a mirror affects the reflected image? This is we are going to find out after performing this experiment.

Material Required

- Hand mirror;
- Pencil;
- Paper;
- 4 books.

Procedure

Support the mirror with the books. Place the paper,under the edge of the mirror. Rest your chin on your hand so that you can see into the mirror, but so that your view of the paper where you will be writing is blocked. Look only into the mirror as you write your name so that it appears correctly in the mirror. Examine your writing.

This is What Happens

Most and maybe all of the letters are upside down.

Science Behind It

Because you are writing so that the letters are correct in the mirror, they are reversed on the paper. Most letters would be written upside down for this experiment except those that are symmetrical (the same on both sides) such as O, E, H, I, and B. These letters look the same with or without the mirror. The mirror gives you a reversed image.

217. Making Siphon

Here's an experiment that doesn't require any work—just as long as you set it up correctly. Then sit back and watch as the physics of flowing takes control.

Material Required

- Two containers;
- A length of plastic hose or aquarium air pump tubing;
- Chair.

Procedure

Fill one container 3/4 full with water. Place this container on the flat seat of a chair. Place the empty container on the floor in front of the chair.

Fill the tubing with water. Stop both ends with your fingers. Place one end of the tubing in the lower container. Place the other end in the upper container.

Remove your fingers from both ends. What happens?

This is What Happens

You've created a siphon. It is a bent tube with arms of unequal length. Liquid travels through the siphon from one container to the other until the levels of liquid in both containers are the same.

Air pressure pushes down on the surface of the liquid that is found in both containers. The force of this downward push supports a column of fluid in the arm directly above this liquid. However, the longer arm contains a greater "weight" of fluid pushing down than the smaller arm. The arm with the greater weight overcomes the upward resistance (due to air pressure) and the liquid slides down this arm. As it moves into the lower container, the liquid draws up fluid through the siphon's shorter arm.

218. Pencil Illusion

The intention of this experiment is to demonstrate the strobe effect of a television picture.

Material Required

- Television;
- Pencil.

Procedure

With the lights on, swing the pencil up and down quickly and observe how it looks. In a darkened room with only the television on, hold the pencil in front of the screen. Quickly swing the pencil up and down 4 to 5 times.

Caution: Never watch television in a room lit only by light from the television screen. It is extremely bad for your eyesight.

This is What Happens

In light, the moving pencil produces a continuous blur. In the dark, separate images of the pencil are seen in different places.

Science Behind It

Instead of a continuous blur of the moving pencil, separate images are seen, because the light from the television screen is not constant. Thirty pictures are flashed on the screen each second. Between the pictures the screen goes black, and the pencil moves to a new place. The movement of the pencil is not seen when there is no light. This blinking light gives the illusion that the pencil is moving in slow motion. You do not notice any blinking while watching television because your eyes retain the image of each picture long enough to receive the flashing of the next picture.

219. Pressure Drop

In this experiment, you'll be punching a pencil through an empty milk carton.

Material Required

- Milk carton;
- Tape;
- Pencil;
- Water;
- Sink.

Procedure

Rinse out an empty milk carton. Use a sharp pencil to carefully poke two holes in the carton. The holes should be positioned one on top of the other and several inches apart.

Use tape to cover up the holes. Fill the carton with water and place it next to a sink so that the openings aim into the sink basin. Pull off the tape. How do the two jets compare? Can you explain your observations?

This is What Happens

To understand the difference in jets, let's take a look at the amount of water that is found above each hole. The hole that is near the container's top has less water above it than the corresponding hole below it.

Less water produces less weight. Less weight results in a lower pressure. Therefore, the jet near the top of the container is under less pressure. Hence, its stream is not as energetic as the jet that emerges from the lower (higher pressure) hole.

220. Diffracting Grating

How a diffracting grating affects light? This is what we are going to find out.

Material Required

- Cotton handkerchief;
- Lamp.

Procedure

Remove the shade from the lamp. Stand about 6 ft. (2 m) from the glowing bulb. Look at the light through the stretched cloth.

This is What Happens

A starburst of light with dim bands of yellow and orange colours appear around the light.

Science Behind It

The cloth acts like a *diffracting grating*, which separates light into its individual colours. Diffracting gratings are made by using a diamond point to cut as many as 12,000 lines per ½ in. (centimetre) on a piece of glass or plastic. The spaces between the woven threads in cloth separate the light, but since the holes in the weave are large, not as many separate colours are seen as one would observe through a diamond cut grating. The starburst pattern is due to the bending or changing direction of the light.

221. Crazy Balloon

Fill a balloon with air and knot it. The balloon stays about the same size because there is a balance of forces. The force exerted by the air inside the balloon equals the force pushing in by the atmosphere. Since things are equal, the balloon stays the same size.

In this experiment, you'll see what happens when forces are unequal.

Material Required

- 1-litre soft drink container;
- Balloon;
- Warm water.

Procedure

Run a stream of warm water over the outside of a 1-litre plastic container. After about a minute remove the container from the stream. Stretch a rubber balloon over the opened mouth of the container and allow it cool. You can hasten the cooling by placing the container in cool water. What happens to the balloon?

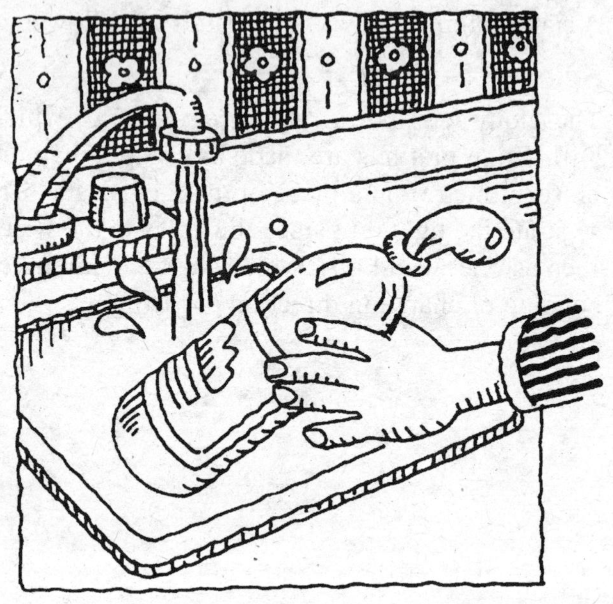

This is What Happens

As you warmed the container, you energized the air particles that were inside the container. When heated, the particles moved faster. As they moved, many of them escaped through the neck of the container. Then the passageway was sealed when you stretch the balloon over it.

As the container cooled, the trapped air particles lost energy. Since the air particles were not as energetic, they moved slower. There were also fewer of them (don't forget that many escaped).

This resulted in a greater concentration of air on the outside than the inside of the container. The difference in concentration produced a similar difference in pressure. This pressure difference pushed the balloon into the container.

222. Cooling Spray

The purpose of this experiments is to determine if gas pressure affects temperature.

Material Required

- Aerosol can of first-aid medicine.

Procedure

Hold the spray can near your arm. Spray some of the contents onto your arm and note the temperature of the spray.

This is What Happens

The spray feels cold.

Science Behind It

The can contains medicine and gas to produce a high pressure. When the nozzle opens, the gas escapes quickly, carrying with it drops of liquid. The cold temperature is due to the rapid expansion of the gas. Increasing the volume of the gas decreases the pressure on the gas and causes its temperature to drop.

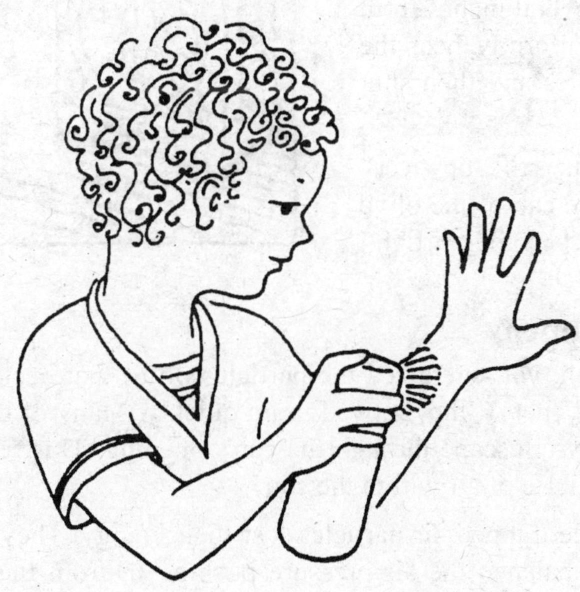

223. Crushing Cans

This is a great experiment to know about air pressure.

Material Required

- Large empty olive oil can (with cap);
- High wattage hair dryer with safety plug;
- Two kitchen oven mitts.

Procedure

Thoroughly wash an empty olive oil can with a soap water solution. Pour out the solution and rinse well. Make sure to wash any oil from the inside of the cap as well. Turn the can upside down and allow it to dry overnight.

Place the clean and dry can on a protective surface far from a sink or running water. Put on a pair of oven mitts. Use a high wattage hair dryer to warm the outside of the can. Keep the nozzle of the dryer several inches from the can's surface. Uniformly heat the can for several minutes and then shut off the dryer.

With the oven mitts still on, very carefully screw on the cap of the olive oil container so that it becomes airtight. Now sit back and watch.

This is What Happens

As you heated the can, you energized the particles of air that occupied the inside space. As these particles gained energy, they moved more quickly. Many of these "energized" particles gained enough energy to escape through the can's opening. This resulted in fewer (although highly energetic) particles of air within the can.

As the can cooled, the trapped air particles lost their energy. They could no longer produce enough collisions to balance the air pressure pushing in from the outside. Since the inner pressure could not counteract the outside air pressure, the can collapsed.

224. Warm Band

In this experiment, we will learn more about conservation of energy. What happens when one form of energy changes in other form of energy? This is what we will find out.

Material Required

- Rubber band.

Procedure

Place the rubber band on your forehead and note the rubber band's temperature. Hold the rubber band between your thumbs and index fingers with your thumbs touching. Stretch the rubber band. Quickly touch the stretched band to your forehead.

Note: Your forehead is sensitive to heat and can therefore be your sensing device.

This is What Happens

The stretched rubber band feels warm.

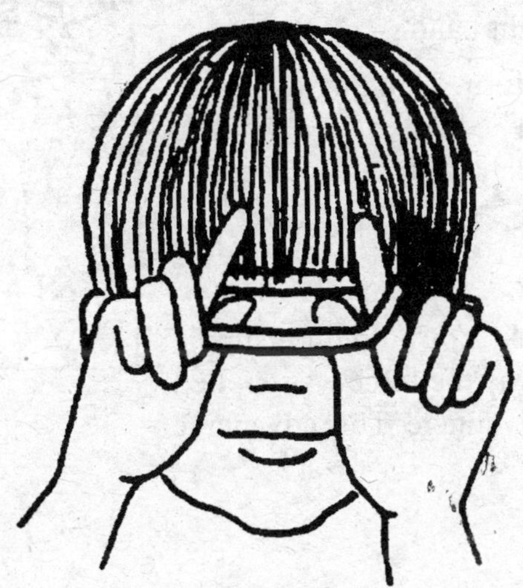

Science Behind It

The rubber band is made of molecules coiled like a spring. Stretching the rubber band straightens the coils; they recoil when the band is released. You used mechanical energy—energy of moving things—to pull the coils of molecules apart, and the rubber band used energy to pull the coils back together. Some of the mechanical energy was changed into heat energy. Energy was needed to stretch the rubber band, and energy was needed to restore it to its original shape. If there were no changes in the molecular structure of the rubber band, the amounts of energy used to stretch and to recoil the rubber band would be the same. The energy changed from one form to another, but it was not lost. This is called conservation of energy.

225. Crank – A Simple Machine

When it comes to do work, it often helps to have an advantage. That's where machines come in. Machines are devices that give us an advantage.

There is a trade-off, though. Machines are designed so that less force is needed to accomplish a task. The lessened force, however, must be applied for a longer time or over a greater distance.

Material Required

- A couple of paper clips.

Procedure

Unbend a paper clip. Straighten it out as best as you can.

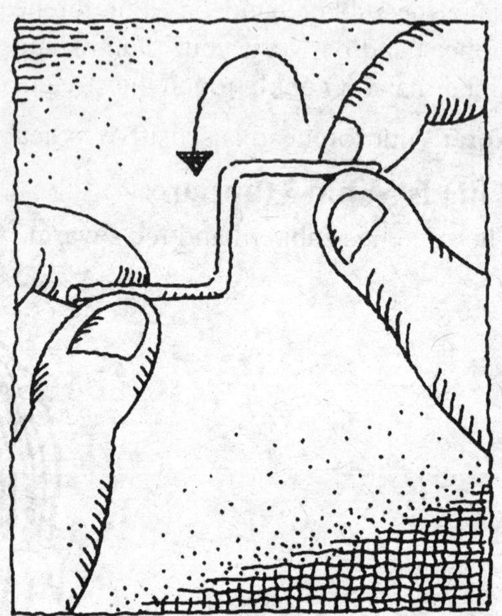

Hold one end of the clip tightly between the thumb and index finger of your left hand. Grasp the free end of the clip with your right hand. Try to roll the clip using the fingers of your right hand.

Most likely, the clip didn't turn. Spinning a straight wire that is held in place can be hard to do. What you need is an advantage—a mechanical advantage!

Place two bends in the clip so it looks like this:

This is What Happens

The clip has become a crank. You've built a simple machine! By changing the way the force was applied to the clip, you gained an advantage. This advantage made turning the clip easier.

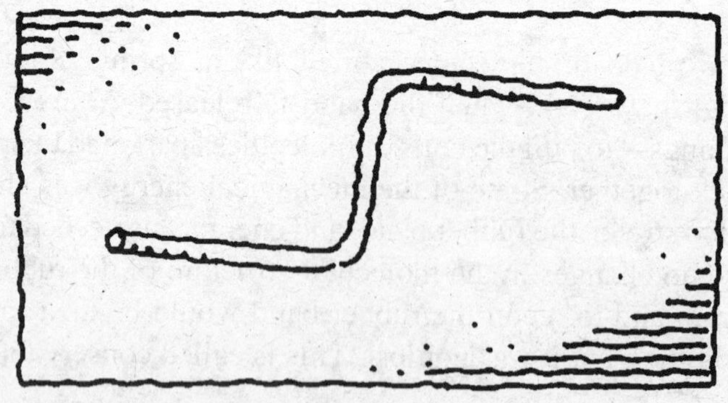

226. Cold Foot

The purpose of this experiment is to demonstrate the conduction of heat energy.

Material Required

- Aluminium foil;
- Small throw rug or rug sample.

Procedure

Cut a piece of foil a little bigger than your foot. Place the foil and the carpet on a tile floor. Allow them to remain undisturbed for 10 minutes. Put one bare foot on the aluminium foil and the other on the carpet. Observe any difference between the feel of the temperature of the aluminum piece and that of the carpet.

This is What Happens

The metal foil feels colder than the carpet.

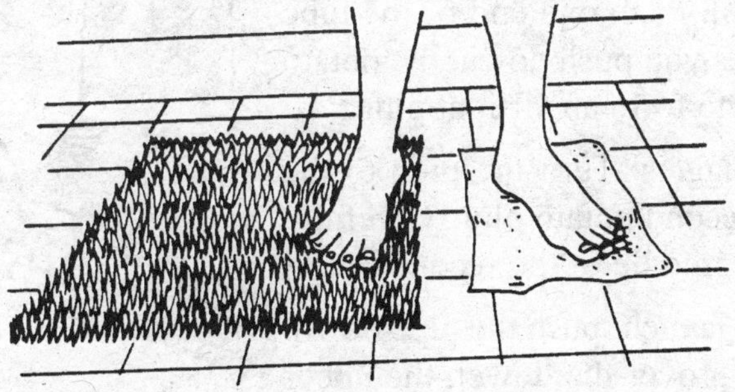

Science Behind It

A good heat conductor allows heat to move through it. This is accomplished by movement of individual molecules. One molecule receives the heat energy and starts bouncing around. During these jerky motions, the molecule bumps into a molecule nearby, causing it to bounce around. This exchange of energy continues through the entire piece of material if it is a conductor of heat. Some materials are very good conductors of heat and others are not. Molecules in metals like aluminium can easily move around and are very good conductors of heat energy. Good heat conductors are also good electrical conductors. The piece of carpet is a poor conductor of heat, which means that its molecules do not move easily. Allowing the foil and carpet to sit for 10 minutes gave both materials time to reach room temperature.

Things feel cold to the touch when heat energy is drawn away from your skin; things feel warm when heat energy is transferred to your skin. The aluminium feels colder than the carpet because it is a good heat conductor and the heat, from your foot starts moving through the metal. The carpet is a poor heat conductor and actually blocks heat loss from your foot.

Poor heat conductors are also called insulators. To keep a house from losing its energy, insulators are placed in the space between the inside and outside walls to block the flow of heat into or out of the house.

227. Potato Missiles

Both the laws of physics and the presence of air come handy when it comes to make potato missiles.

Material Required

- 1-2-foot length of 1-inch tube or pipe (made from metal or PVC material);
- Raw potato;
- 1 -foot wooden dowel (less than 1 inch in diameter);
- Knife;
- Safety goggles.

Procedure

Cut several slices of potato. The slices should be larger than the diameter of the tube and about as thick as your thumbnail. Push one end of the tube onto the potato slice. As you push down, the potato should fill the opening to form an airtight plug.

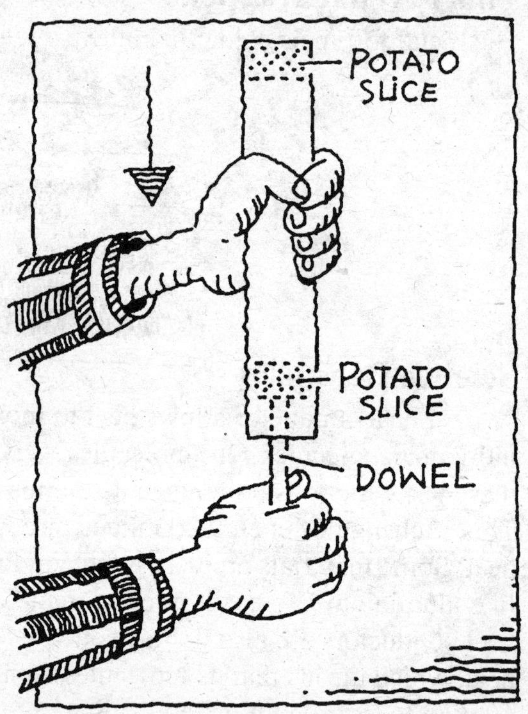

Put, on your safety goggles. Turn the pipe to the other end and form a second potato plug. Carefully position a dowel in the middle of the second plug.

When you are ready to launch, push the pipe down. As the pipe slides down over the dowel, the upper potato plug will be launched skyward.

This is What Happens

As you lower the pipe onto the dowel, the bottom potato plug rises. As it moves up the pipe, it compresses the air trapped within the tube.

Eventually, the pressure becomes great enough to push the upper potato plug from the pipe. The plug is propelled by a blast of compressed air and "pops" out of the pipe.

228. Popping Popcorns

We all love to eat hot popcorns. But have you ever wondered why popcorn pops.

Material Required

- Unpopped popcorn;
- Hot-air popcorn popper.

Procedure

Observe the shape and size of a few unpopped corn kernels. Ask an adult helper to assist you in setting up the popcorn popper. Observe the shape and size of the corn kernels as they are being heated.

This is What Happens

The corn kernels change from small, hard, orange, kernels to large, soft, white, ball-shaped structures.

Science Behind It

The tough outside of the unpopped kernel is called the pericarp. This is the part that often gets stuck in your teeth when you eat popcorn. The inside is filled with starch that expands into the white, fluffy popcorn. The small amount of water inside the kernel makes the explosion possible. As the kernel is heated, the liquid water evapourates —changes to a gas. The gas expands and pushes so hard on the pericarp that it breaks and the starch tissue inside is blown outward. The pop noise is the sound of steam escaping and the pericarp breaking.

229. Racing Spool

Rubber bands are made from an elastic material. When this material is stretched, it stores energy. The more twists in the band, the greater the amount of stored energy. When the band is released, the stored energy is given off. The following device harnesses the energy released by unwinding elastic to power a racing spool.

Material Required

- Empty thread spool;
- Cotton swab;
- Metal washer;
- Rubber band;
- Paper clip;
- Tape.

Procedure

Insert a rubber band through the centre hole of a thread spool. Slide one end of the rubber band over a paper clip. Use tape to secure the clip to the side of the spool.

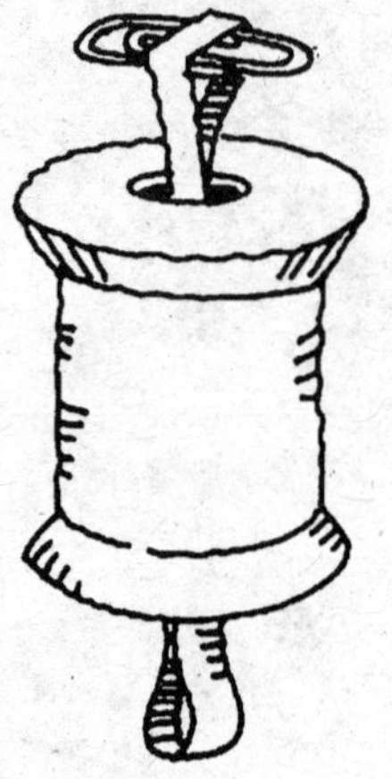

At the other end, insert the rubber band loop through a metal washer. Then insert a cotton swab into the open loop as shown in the illustration.

Wind the swab so that the rubber band twists. Turn the spool on its side and place it on a flat surface. Release. What happens? How can you control the path of this racer?

This is What Happens

The twists of the rubber band store energy. When the spool is released, these twists are allowed to unwind. The unwinding motion is transferred to the spool through the fixed paper clip. This causes the spool to spin and race along the floor.

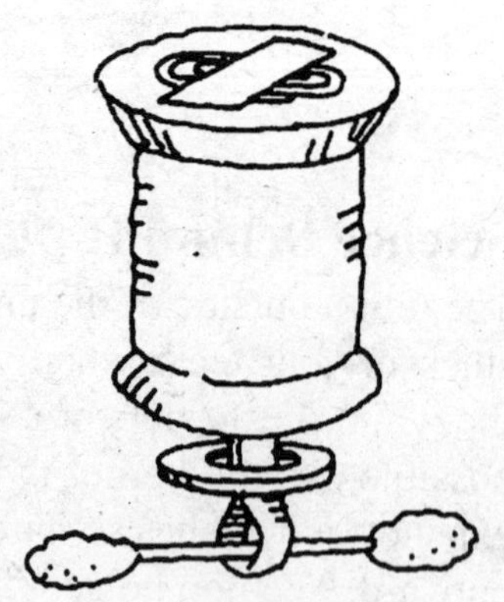

230. Bouncer

We all know that a ball bounces when thrown. Can we change the rate of ball bouncing? or does the temperature have any effects on the bounce of a rubber ball. Let's find out in this experiment.

Material Required

- Tennis ball;
- Yardstick (metre stick);
- Refrigerator with freezer.

Procedure

Hold the yardstick with one hand and place the ball at the top edge of the yardstick. Release the ball, and observe the height of the first bounce. Repeat three times to get an average of the bouncing height.

Place the ball in a freezer for 30 minutes. Again measure the height that the ball bounces when released from the top of the yardstick.

This is What Happens

The ball does not bounce as high when it is cold.

Science Behind It

Rubber is made of thousands of small molecules joined to form long chains. At room temperature, the chains easily push together and pull apart to make the ball bounce. The chains of molecules become rigid when the ball is chilled. The flexibility of the chains of molecules allows the tennis ball to bounce. Playing tennis in cold weather would affect your game.

231. Can on Command

Here's one of those strange tricks that can baffle your audience. Roll a can and watch as it travels along the floor. Then whistle for it to return. As if responding to the command, the can stops moving and spins back to you. Magic? No, science.

Material Required

- Empty coffee can;
- Metal twist tie;
- Two plastic coffee can covers;
- Several long rubber bands;
- 1-5 ounce fishing sinker;
- Nail.

Procedure

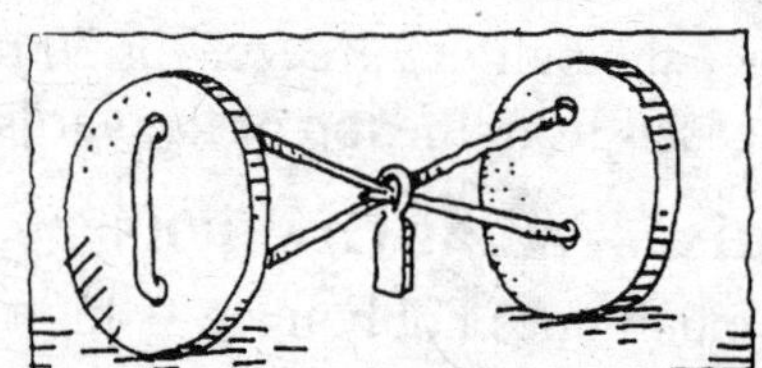

Use the nail to carefully punch two holes into each of the coffee can lids as shown in the illustration. Then cut the rubber bands to form lengths of elastic. Tie the ends of the elastic together to form one strand of rubber about 20 inches long.

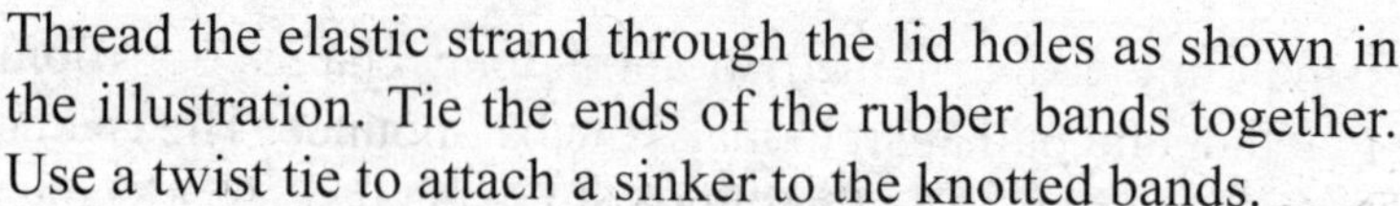

Thread the elastic strand through the lid holes as shown in the illustration. Tie the ends of the rubber bands together. Use a twist tie to attach a sinker to the knotted bands.

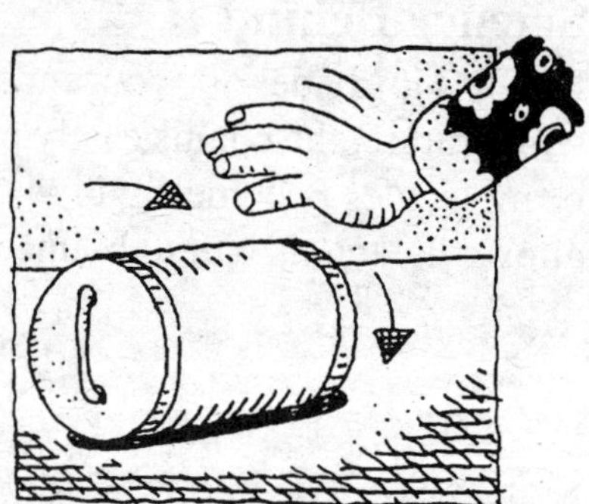

Have an adult fold one of the lids and push it through the metal-container. Place the lids on the opposite ends of the can. Roll the can and watch as it listens to your "come on back" command.

This is What Happens

As the can rolls, the fishing weight doesn't spin. Instead, it acts as a stationary anchor. As the can spins, the elastics twist up against the fishing weight. By the time the can stops, a considerable amount of energy has been stored in the rubber band twists. When the strand unwinds, the released energy is changed into the motion of the can's return.

232. Boom!

The purpose of this experiment is to show the effect of solids on the speed of sound.

Material Required

- Clear plastic drinking glass;
- Rubber band.

Procedure

Stretch the rubber band around the glass as shown in the diagram. Hold the bottom of the glass against your ear. Gently strum the stretched rubber band.

This is What Happens

The sound heard is very loud.

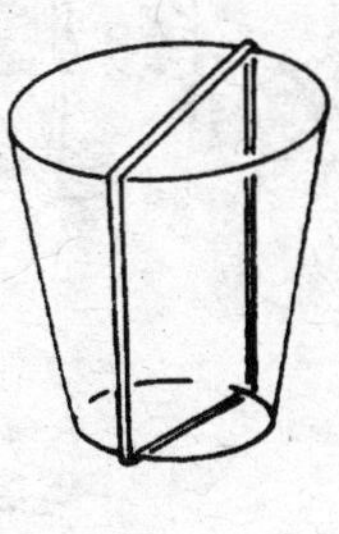

Science Behind It

Sound is produced when objects vibrate. As the object moves back and forth, it hits against air and any other object near enough to be touched. When vibrations start air moving, the continuous ocean of air around you transfers the energy to your ears, and you register that sound has been produced. Vibrations move much more slowly through the air—a gas—than they do through liquids or solids. The vibrating rubber band causes the air around it to move, but the booming sound that you hear is because the solid plastic transmits the vibrations to your ear.

233. Propeller Power

Do a propeller and a wood screw have in common? Plenty. They are both simple machines that use something called an inclined plane. An inclined plane is a simple machine that produces an advantage. Perhaps the most familiar inclined plane is your plain old ramp.

Now imagine that ramp wrapped around a central core. Presto, a screw. Look at the sides of the screw and you'll see the ramp begin at the screw's tip and circle its way back along the screw's shaft. A propeller also uses this type of wrapped ramp. In fact, propellers are known as "air screws."

Material Required

- Scrap block of foam packing;
- Two paper clips;
- Tape;
- Modeling knife;
- Kite string;
- Rubber band;
- A toy propeller assembly (removed from a cheap and broken wooden toy);
- Safety goggles.

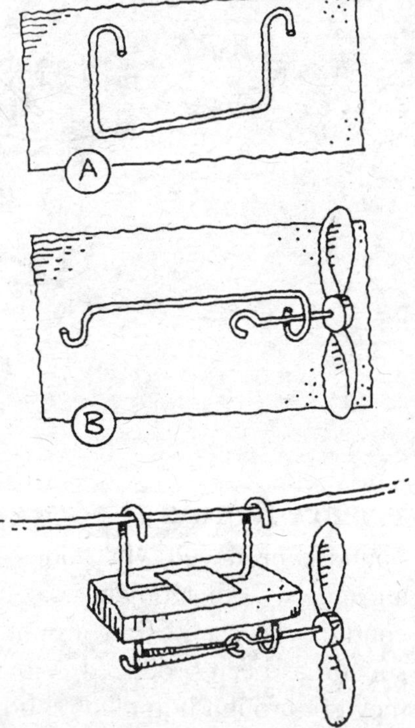

Procedure

Open and rebend a paper clip so that its ends form two similar loops (clip A). Use tape to anchor this clip to the top of the foam block.

Open and bend a second paper clip (clip B) so that the loop at the back of this piece will hold the propeller assembly. Use tape to attach this clip to the bottom of the foam block. Stretch a rubber band from the front hook to the hook on the propeller shaft.

Tie a kite string across a room. Put on your safety goggles. Wind up the elastic. Hang your craft from the two top hooks. Release the wound propeller and watch your craft shoot along the string!

This is What Happens

As the elastic unwinds, it transfers its motion to the propeller. The spinning propeller cuts through the air. As it rotates, it produces a substantial flow of air. This flow produces a difference in air pressure that causes the craft to move ahead.

234. Bottle Organ or Jal Trang

How frequency affects the pitch of sound? Let's try this musical experiment and find out.

Material Required

- 6 small-mouthed bottles of comparable size;
- Metal spoon.

Procedure

Pour different amounts of water in each bottle. Gently tap each bottle with the metal spoon. Note the difference in the pitch produced.

This is What Happens

The bottle with the most water has the lowest pitch.

Science Behind It

Sounds are made by vibrating objects. The number of times the object vibrates—moves back and forth—is called the frequency of the sound. As the frequency increases, the pitch of the sound gets higher. Tapping on the bottles causes the bottles to vibrate. As the height of the water increases, the pitch of the sound gets lower.

235. Balloon-powered Cars

This is your chance to design your own balloon-powered racer. Go crazy with the design.

Just think, there are no restrictions. Your balloon-powered car can have four wheels like a car or it can have three wheels like a tricycle. You can build the wheels out of cardboard, thread spools, bottle tops, or even use wheels that you've ripped off from toy cars. It's up to you.

Material Required

- Balloon;
- Tape;
- Cardboard;
- Thread spools;
- Pair of scissors;
- Glue;
- Rubber bands;
- Wooden sticks;
- Tape;
- Toy wheels;
- Straw;
- Whatever other materials you want to use;
- Safety goggles.

Procedure

Think about your design. You may want to brainstorm different ideas so that you won't be locked into a finished "look" right away. Take your time. Try drawing several designs. You may even want to draw up a set of blueprints.

Once you have decided on a design, try building it. Don't worry if you get into problems with its construction. Just figure out a way to get around the obstacle. If there are too many problems with your design, you can always try thinking up a new car. So don't get hung up on what you don't have. Be creative with the materials you do have!

This is What Happens

It's Newton's Third Law combined with some creative freedom and mixed with a good dose of heavy-duty gray cells.

236. Singing Glass

The aim of this experiment is to demonstrate how friction can cause a glass to vi-brate.

Material Required

- Stemmed glassware (this will work better if the glass is very thin);
- Dish detergent;
- Sink or large pan;
- Vinegar;
- Small shallow bowl.

Procedure

Use the dish detergent to make a warm, soapy water solution in a sink or large pan. Wash the glass and your hands in the warm, soapy water, and rinse well. Place the glass on a table.

Pour a thin layer of vinegar into the small bowl. Hold the base of the glass against the table with your hand. Wet the index finger of your free hand with vinegar and gently rub your wet finger around the rim of the glass.

This is What Happens

The glass starts to sing when its rim is rubbed.

Science Behind It

Washing the glass and your hands removes any oil that might act as a lubricant. The vinegar also dissolves any oil that might be present and increases the friction between your skin and the glass. Rubbing your finger around the rim causes the glass to vibrate because your finger skips and pulls at the glass. This irregular touching on the glass rim is like tiny taps that cause the glass to vibrate. The air inside the glass is struck by the vibrating glass and begins to move back and forth in rhythm with the glass. The pitch of the sound you hear is due to frequency (number of vibrations per second).

237. Rainbow in a Tray

White light is more than meets the eye. It's a mix of all the colours of the rainbow (red, orange, yellow, green, blue, and violet). These colours make up the visible spectrum. There are several ways that white light can be separated into its component colours. Here's one:

Material Required

- Small mirror;
- Tray;
- Water;
- Sheet of white paper.

Procedure

Fill a tray with water. Set the tray in bright sunlight. Position a mirror inside at one end of the tray. The mirror should be set at an angle that is supported by the tray edge.

Look at the mirror's reflection that is cast upon a nearby surface. What do you see? To make the pattern appear brighter, place a sheet of white paper beneath its reflection.

This is What Happens

Light travels in waves. Like ocean waves, it has high points called crests and low points called troughs. The distance from one crest to the next is a measurement called wavelength.

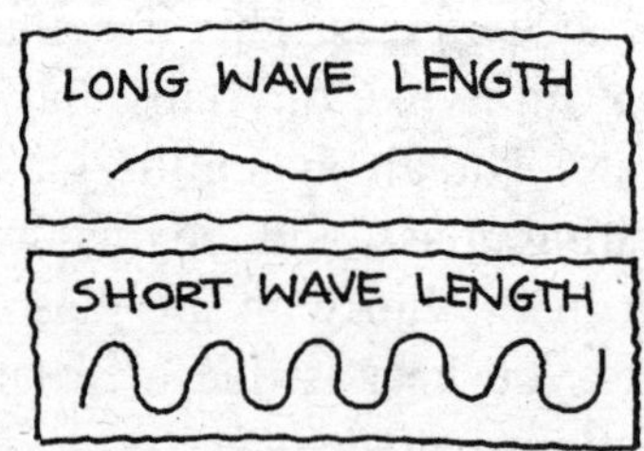

A beam of white light contains rays of many different wavelengths. Each wavelength has its own colour. Red has the longest wavelength. Next comes orange, followed by yellow, green, and blue. Violet is the colour with the shortest wavelength.

When white light is reflected through the mirror and water, it is separated into its component colours. The colours spread out as a side-by-side pattern is called a spectrum.

238. Musical String

This experiment demonstrates factors that affect the pitch stringed instruments.

Material Required

- 2 buckets, 1 qt. (1 litre) is large enough;
- Rocks, enough to fill both buckets;
- 2 pencils;
- String;
- Scissors;
- Table, about 3 ft. (1 m) wide.

Procedure

Cut a length of string twice as wide as the table. Place the string across the table and tie the ends to the bucket handles so that the buckets hang freely. Place the pencils on the edge of the table under the string.

Fill each bucket half full with rocks. Pluck the centre of the string with your fingers. Move the pencils closer together and again pluck the centre of the string.

Fill the buckets with rocks, move the pencils to different positions, and pluck the string at each different pencil position.

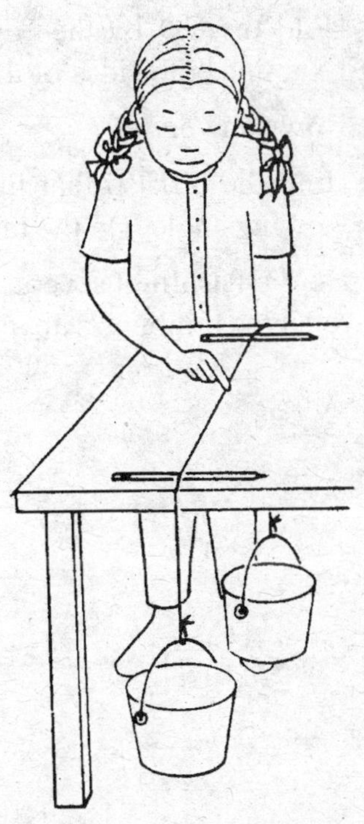

This is What Happens

Adding rocks to the buckets and moving the pencils closer together produces a higher pitched sound when the string is plucked.

Science Behind It

The pitch of a vibrating string obeys certain laws. Length and tension are two factors that affect the pitch of sound produced by a stringed instrument. The pitch is a result of frequency (the number of vibrations produced per second). The faster the string vibrates, the higher is the pitch of the sound produced. The string vibrated faster as the tightness of the string or tension was increased by adding light to the buckets. Moving the pencils closer together shortened the string that vibrated. The shorter the string, the faster it vibrated. Increasing the tension and shortening the length of a plucked string produces a higher pitched sound because the string vibrates faster. This is what happens when a guitar string is held down with one hand while being plucked with the other. Violins, cellos, and other instruments with strings work the same way.

239. Colours of CD

One colour stores sounds. The other stores programs (which often include sounds). But even though their data is different, music and computer CDs are very similar. Both disks are storage mediums. They contain a huge amount of information stored in a code of tiny pits. And even though you can't see these pits, you can observe the great optical effects they produce!

Material Required

- A compact disk;
- Flashlight.

Procedure

Hold the CD by the edges of the disk (otherwise, you might scratch or damage the data). Examine both sides of the disk. Which side contains the stored information?

How can you tell?

Hold the disk so that light reflects off its surface. What do you see? Gradually change the viewing angle. Do the rainbows change? Does the CD cast a rainbow shadow on a wall?

Hold a flashlight several feet from the CD. Aim the beam on the shiny surface. Can you find a reflected rainbow? Move the beam closer to the CD. Can you uncover any rainbow reflections?

This is What Happens

The shiny surface of the CD contains an incredible number of closely spaced pits. The spacing is so close that their pattern can affect light. When white light falls on this surface, the colours separate out. The order of separation depends upon the colour's wavelength. Colours with the longest wavelengths (red) bend the least. Those with shorter wavelengths (violet) bend the most. This bending produces spread-out bands of the spectrum.

240. Ruler Sounds

The purpose of this experiment is to demonstrate the effect that length has on the sound of a vibrating material.

Material Required

- Ruler;
- Table.

Procedure

Place the ruler on a table with about 10 in. (25 cm) of the ruler hanging over the edge of the table. Press the end of the ruler against the table with your hand. With your other hand, push the free end of the ruler down and then quickly release it. As the ruler moves, slide it quickly onto the table. Observe the sounds produced.

This is What Happens

The sound produced changes from a low to a high pitch as the length of the ruler extending over the edge of the table decreases.

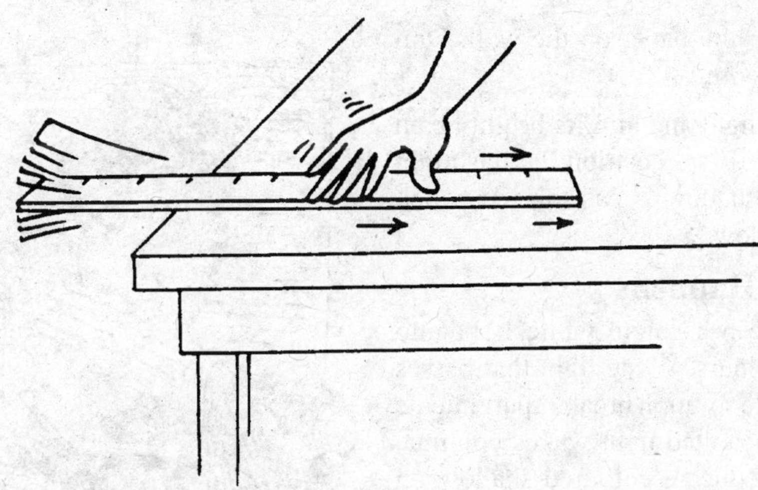

Science Behind It

Sound is produced by vibrating materials. The pitch of the sound becomes higher as the number of vibrations increases. The longer the vibrating material, the slower the up-and-down movement and the lower the sound produced. Shortening the ruler causes it to move up and down very quickly, producing a higher pitched sound.

241. Bubble Spectrum

Have you ever looked at a film of oil floating on water? If so, did you see a swirling rainbow? The oil, like a mirror in water, separated white light into its component colours. In this next experiment, you'll replace the oil with bubble solution to get some colourful results.

Material Required

- Bubble solution;
- Small dish;
- Flashlight;
- Pipe cleaner;
- Tape;
- Sheet of white paper.

Procedure

Bend a pipe cleaner into a loop. Make sure to keep a small handle for holding the loop. Fill a dish with bubble solution.

Dip the loop into the bubble solution. Let the solution saturate the loop. After several minutes, gently remove the loop. What do you see? Are any colours present?

Tape a sheet of white paper on the wall. Dim the lights in the room.

Turn on the flashlight and aim its bright beam through the soapy loop. Position the flashlight so that the loop shadow is cast on the paper. Describe the shadow.

This is What Happens

Very thin films can separate white light into its component colours. White light that passes through the bubble solution breaks apart into its spectrum. The separated light waves continue travelling and produce a coloured shadow on the nearby wall.

242. Straw Flute

The aim of this experiment is to find out if the length of a flute affects the pitch of the sound it produces.

Material Required

- Drinking straw;
- Scissors;
- Ruler;
- Marking pen.

Procedure

Make a 1/2 in. (1.3 cm) cut on each side of the straw's end. This forms the reed part of the flute. Place the reed in your mouth. Push on the reed with your lips and blow. You may have to try several times and change the pressure of your lips in order to produce a sound. As you play the straw flute, cut the end of the straw off with the scissors and observe any change in pitch.

This is What Happens

The pitch of the sound gets higher as the length of the straw decreases.

Science Behind It

The sound produced is due to the vibration of the straw and the air inside it. The longer the column of vibrating air inside the tube, the lower is the pitch of the sound.

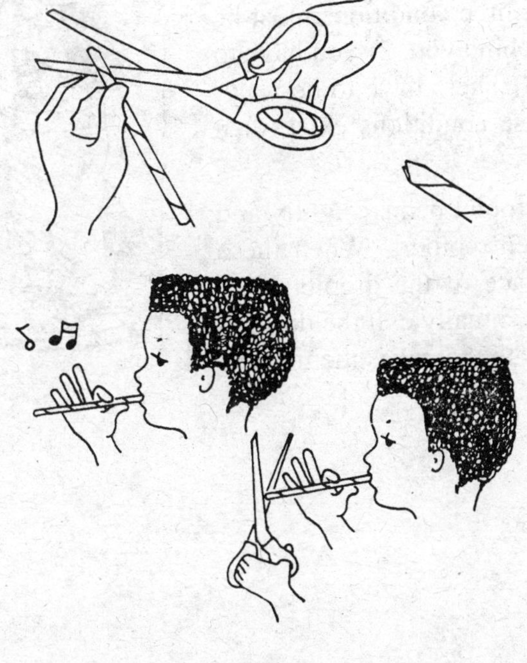

243. Rainbows in Garden

Have you ever shot the spray of a garden hose straight into the air? If the lighting was right, a rainbow may have appeared in the falling mist. Think back to the conditions needed to make this rainbow. Was the sun in front of you or behind you? In this experiment, you'll once again use a water mist to create a rainbow. Your water source, however, will be limited to a spray bottle.

Material Required

- Water bottle with spray top water;
- Bright light.

Procedure

Fill a water spray bottle with water. Stand outside on a bright sunny day and face away from the sun.

Hold the spray bottle in front of you. Aim high and pump out a continual mist of water. You may have to adjust the nozzle so that a fine mist is released.

Look into the falling mist. What do you see? How much of a rainbow can you observe? What colour is on the inner side of the arc? What colour is on the outermost side?

This is What Happens

In order to see a rainbow, three conditions must be met: 1) The sun must be behind you. 2) You have to look away from the sun. 3) There have to be water droplets in the sky. If these conditions exist, then rainbows can appear.

Light that enters a water droplet bends slightly and separates into its component colours. When these colours strike the far surface of the droplet, they get reflected. When they eventually exit the droplet, the colours follow a final path that takes them to the observer's eye.

244. Clucking Chicken

In this simple experiment, we will use a vibrating string to produce a clucking chicken sound.

Material Required

- Paper cup, 6.4 oz. (192 ml);
- Kite string, 24 in. (60 cm);
- Sponge, rectangular kitchen sponge;
- Pencil;
- Toothpick;
- Water.

Procedure

Use the pencil to punch two holes about ½ in. (1.5 cm) apart in the bottom of the cup. Push the string through the holes and tie it on the outside of the cup. Insert the end of the string in one of the holes and pull it through so that the string hangs out of the cup. Place a toothpick under the loop of string on the outside of the cup with the ends of the toothpick extending over the edges of the cup.

Cut a 1 in. x ½ in. (2.5 cm x 1.3 cm) section from the sponge. Tie the end of the string around the centre of the piece of sponge. Wet the sponge with water. Wrap the wet sponge around the top of the string. Squeeze the sponge against the string as you move the sponge down the string using jerky movements.

This is What Happens

A sound is produced like that of a clucking chicken.

Science Behind It

The water allows the sponge to move down the string, but there is enough friction to cause the string to vibrate, because the sponge skips and pulls at the string. This irregular touching on the string produces tiny taps that force the string's molecules to move back and forth. The vibrating string strikes the molecules in the cup, and the cup's molecules strike the air molecules, causing them to move back and forth in rhythm with the cup and string. The pitch of the sound is due to frequency.

245. Mixing Colours

Every time you observe white, all of the colours of the spectrum are present. Is this True or False?

Would you be surprised to learn that the answer is false? Your brain can be tricked into seeing white. This deception occurs when you view the right mix of certain colours and shades. Two colour combinations, such as blue mixed with yellow or red mixed with bluish-green (called cyan), can also appear white.

Can other combinations produce the illusion of white? Find out in this easy experiment.

Material Required

- Three flashlights;
- Transparent plastic report covers (red, blue, and green);
- Pair of scissors;
- Tape.

Procedure

Obtain a red, blue, and green plastic report cover. Cut out a disk from each cover about the size of the flashlight lens.

Tape one disk over the clear lens of each flashlight. Aim the red beam at a nearby white wall. What colour do you see?

While the red remains on, aim the beam of blue so that it overlaps half of the red spot. What happens?

Now aim the beam of green so that all three colours fall on the same spot. What happens as the colours are added to each other?

This is What Happens

When light beams fall upon each other, the combined colours are added together. Unlike the mixing of paint pigments, this additive process produces lighter colours. When the right combination is produced, we "see" white.

246. Phonograph Speaker

In this experiment, we will make a phonograph speaker with things around us.

Material Required

- Poster board;
- Wrapping paper;
- 33-RPM record —one that is disposable;
- Scissors;
- Masking tape;
- 4 straight pins;
- Record player.

Procedure

Make a large and a small cone from both the poster board and the wrapping paper by rolling and taping as shown in the diagram. These four cones will be used as megaphones.

Place a straight pin through the end of each of the four cones. Turn on the record player. In turn, hold each of the cones so that the pin rests on the grooves of the turning record. Observe any sounds produced.

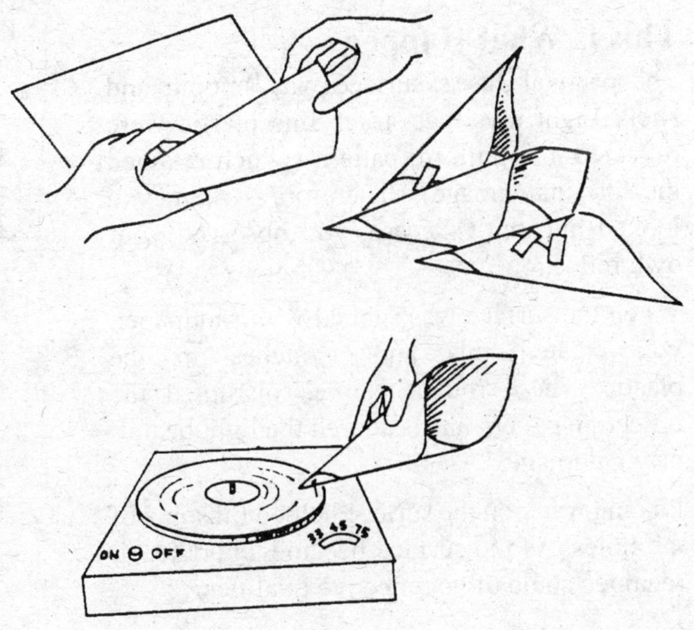

This is What Happens

Music is heard. The loudest and clearest sound was heard through the megaphone made of large, thin paper.

Science Behind It

Sound is caused by vibrating air that reaches your ear. The pin moves up and down in the record grooves and hits against the paper megaphone. The megaphone vibrates the air inside it. It is this moving air that reaches your ear. The thinner paper works best because it is more easily moved. The larger cone gives more paper to vibrate and thus increases the volume of the sound.

247. Scratch Another Mirror

Throw a ball against a smooth wall. Most likely, you'll be able to predict the location of the ball's return bounce. Suppose you threw the ball against a rough or jagged wall. Would it be as easy to guess the spot of this return bounce? Why?

Material Required

- Discarded plastic CD case;
- Sandpaper;
- Water.

Procedure

Look into the clear plastic of a discarded (and soon-to-be-destroyed) CD case. What do you see? What causes the surface of the plastic to produce this reflection?

Take a piece of sandpaper and rough up the surface of the CD. Look into the plastic again. What happened to the reflection?

Wet the rough surface with water. Does the reflective quality of the plastic change?

This is What Happens

The original case's surface was smooth and slick. Light rays that struck this plastic were reflected in a uniform pattern (which retained the original image). Therefore, when you looked into the CD case, you observed your own reflection.

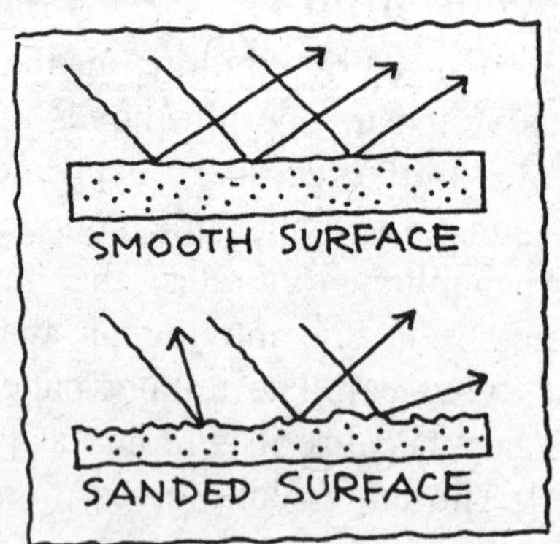

When the surface was rubbed with sandpaper, you created pits and scratches in the plastic. These rough features disrupted the reflection pattern and scattered the light beam's original image.

Placing water on the surface helped fill some of the holes. As the surface became smoother, it regained some of its reflective qualities.

248. Spoon Bell

The purpose of this experiment is to demonstrate how the pitch of sound can be changed.

Material Required

- Metal spoon;
- Kite string, 30 in. (75 cm).

Procedure

Tie the handle of the spoon in the centre of the string. Wrap the ends of the string around your index fingers. Be sure that both strings are of the same length. Place the tips of your index fingers in each ear.

Lean over so that the spoon hangs freely and tap the spoon against the side of a table. Shorten the strings by wrapping more of the string around your fingers. Again, keep the strings the same length. Lean over again and tap the hanging spoon against the side of the table.

This is What Happens

A sound like a church bell is heard. The sound has a lower pitch with the longer string and a higher pitch with the shorter string.

Science Behind It

The metal in the spoon starts to vibrate when struck. Vibration means that something moves back and forth. In this case, the molecules in the spoon move back and forth and hit against each other. When molecules collide, energy is transferred from one molecule to another. The vibrating molecules in the spoon hit against the string's molecules, and the energy is transferred up the string to your ears. The difference in the pitch of the sound is due to frequency (the number of vibrations that reach the ear in one second). High-pitched sounds are produced when many vibrations reach the ear in one second, and the sound gets lower in pitch as the number of vibrations decreases.

249. Multiple Images

Bathrooms are often great laboratories for exploring the properties of mirrors. The sink mirror reflects its image onto the door mirror. The door mirror reflects its image onto the shaving mirror. The shaving mirror reflects its image onto the make-up mirror. Pretty soon, you're not alone. The bathroom is now packed with thousands of people who look exactly like you all trying to do science experiments!

Material Required

- Two mirrors;
- Coin;
- Small toy;
- Clay.

Procedure

Set two mirrors in a lump of clay. Position the mirrors so that they are about 1/2 foot apart. Place a small toy between the two mirrors. Look into one mirror and count the number of toy images. Guess how many toy images will appear in the other mirror. Once you've made your guess, check it out. Do both mirrors have the same image or does one show the toy slightly closer? Why?

Explore how tilting the mirrors are and how changing the distance between them affect the reflections.

This is What Happens

The parallel mirrors create a pattern of reflection that produces an infinite number of reflected images. Each time the image was reflected, its light rays travelled back across the gap to the other mirror. There, this slightly smaller image was reflected back to the other mirror. Again and again and again and again a parade of shrinking images form!

Contents of Vol. 2

Contents of Vol. 3

Contents of Vol. 4

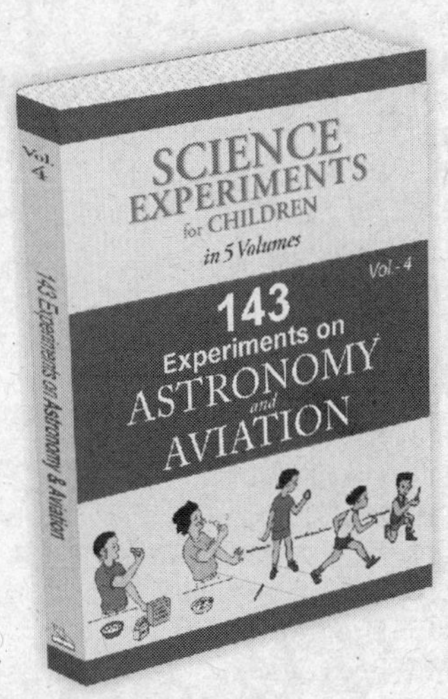

Contents of Vol. 5

POPULAR SCIENCE

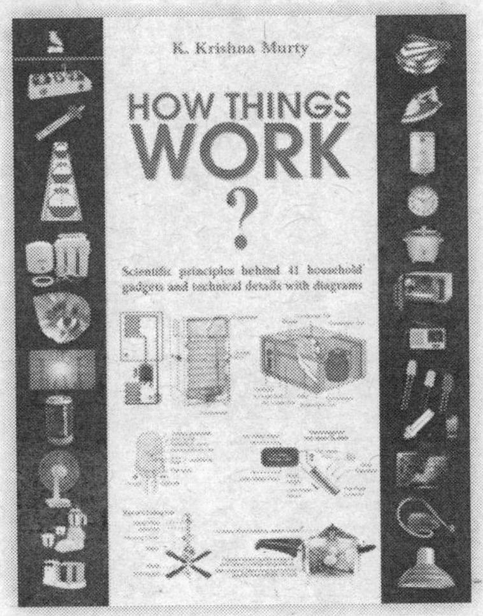

CHILD DEVELOPMENT

MEMORY & MIND DEVELOPMENT

Also available in Tamil.